Searching for ORDER IN the COMPLEXITY of Evolving Worlds

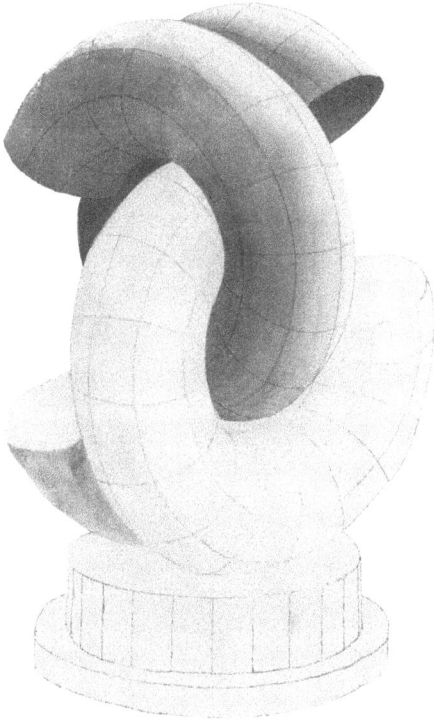

ACKNOWLEDGMENTS

The SFI Press is supported by William H. Miller and the Miller Omega Program.

[THE COMPLEX WORLD]

An Introduction to the
Foundations of Complexity Science

DAVID C. KRAKAUER

ᴴ⸵PR🌿SS

THE SANTA FE INSTITUTE PRESS

1399 Hyde Park Road
Santa Fe, New Mexico 87501

The Complex World:
An Introduction to the
Foundations of Complexity Science
ISBN (HARDCOVER): 978-1-947864-60-3

The SFI Press is generously supported by
the Miller Omega Program.

This volume is a standalone edition of the
introduction to *Foundational Papers in Complexity*
Science, originally published in Volume I.

www.foundationalpapersincomplexityscience.org

EVEN IF ANY GIVEN TERMINOLOGY is a *reflection* of reality, by its very nature as a terminology it must be a *selection* of reality; and to this extent it must function also as a *deflection* of reality.

—KENNETH BURKE
Language as Symbolic Action (1966)

Table of Contents

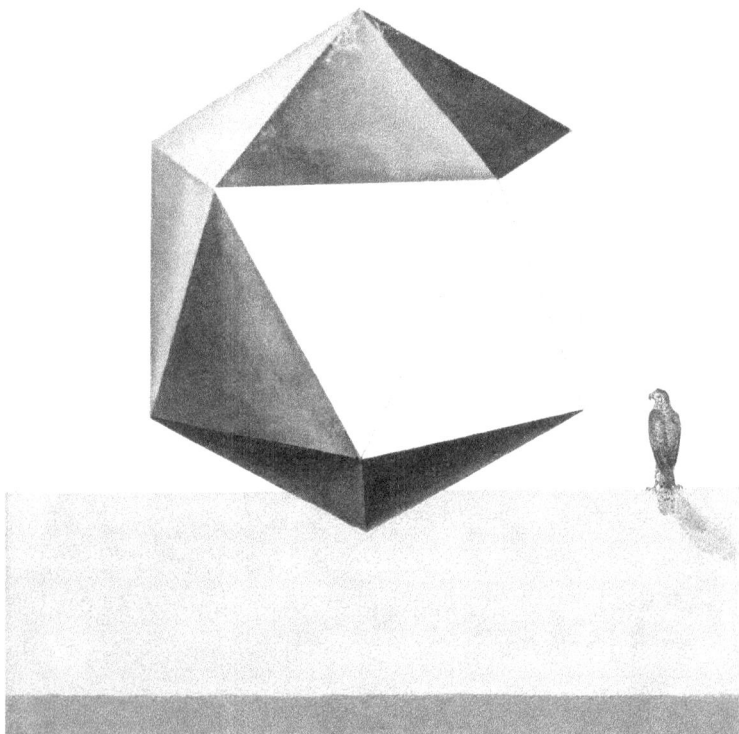

Introduction

Two World Systems

The scientific and social implications of differences between **(A)** closed, reversible, symmetry-dominated, and predictable classical domains and **(B)** open, self-organizing, dissipative, uncertain, and adaptive domains are the subject of this book.[1]

At one limit—**A**—are fundamental regularities described using minimal assumptions, simple rule system, and few initial conditions. At the other—**B**—emergent regularities constructed from contingent histories, described with coarse-grained rules and nested boundary conditions.

Between **A** and **B** there is an uneven spectrum, shaped like a dumbbell with equilibrium structures at one end, non-equilibrium forms of self-organization in the middle, and fully adaptive self-synthesizing organizations with long evolutionary histories at the other.

Analyzing the connections between the simple **A** and the complex **B** requires much more than a powerful measurement device. Here there exist differences that can only be resolved by principles, models, and theories. Interestingly, the more powerful the device—the more finely grained the measurement—the less easily **B** can be distinguished from **A**. Hence

---

[1] *Open* implies positioned within a chemical or electrochemical gradient; *self-organizing* refers to rules amplifying fluctuations into coordinated space-time patterns; *dissipative* refers to dynamics that are irreversible and that break time-reversal symmetry; *uncertain* describes an open-ended space of possible states not enumerable in initial conditions; and *adaptive* describes the capture of information by a system from the environment that promotes persistence and increases multiplicity.

reductionism in the units of analysis not only *fails to explain complexity; it fails to detect it.*

It is not possible to describe differences between **A** and **B** in terms of the fundamental laws of physics and chemistry. Both obey these laws. There is no new physics in a replicating virus not already found in a crystallizing mineral. Indeed, viruses exploit properties of physics far-from-equilibrium to self-assemble within the cell. Human brains are no more or less

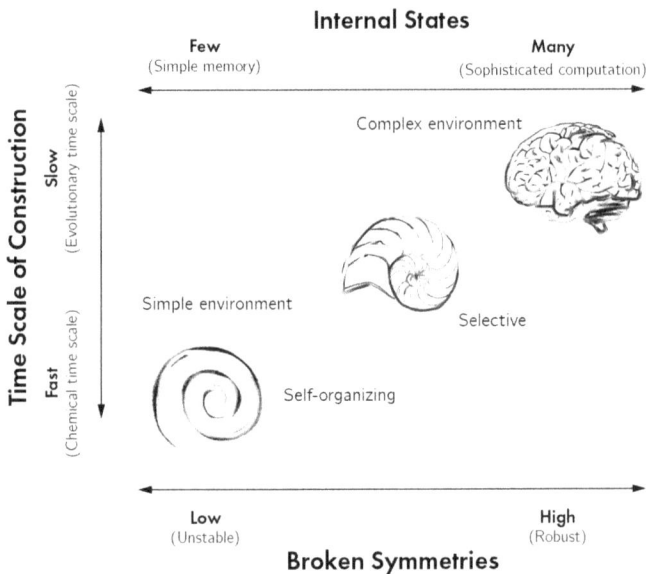

Figure 1. The complex domain: spanning rapidly established self-organizing patterns with few degrees of freedom and slowly evolving lineages with robust information storage. All complex systems emerge first from spontaneous broken symmetries. Simple emergent patterns remain sensitive to changes in boundary conditions. Evolved lineages accumulate a considerable number of metabolically protected "frozen accidents"—intergenerational information-bearing degrees of freedom. These are used to encode adaptive schema (computations) thereby achieving a high degree of autonomy from environmental fluctuations.

concentrations of particles than cannon balls: both depend on covalent chemistry, and both respect the law of gravity. Yet meaningful theories of brains bear no resemblance to fundamental theory in physics.

And the complex world defies many of our best normative intentions. By working along the **A–B** spectrum, engineers have built ingenious electromagnetic communications networks that coordinate to span the globe. But when connecting diverse communities of different basic beliefs and values, the challenging dynamics of **B** often lead to outcomes described as unstable and iniquitous.

Prequel

In 1632 Galileo Galilei wrote of two world systems: the Ptolemaic (geocentric) and the Copernican (heliocentric) (Galilei 2001). Through a series of dialogues, one gradually comes to understand the magnitude of the disruption to metaphysics and empirical knowledge the replacement of Ptolemy will entail. Galileo, played by his alter ego Salviati, nevertheless perseveres in the face of opposition from Simplicio (Galileo 1632, 371):

> **SIMPLICIO**. *The first and greatest difficulty is the repugnance and incompatibility between being at the center and being distant from it. For if the terrestrial globe must move in a year around the circumference of a circle—that is, around the zodiac—it is impossible for it at the same time to be in the center of the zodiac. But the earth is at that center, as is proved in many ways by Aristotle, Ptolemy, and others.*

> **SALVIATI**. *Very well argued. There can be no doubt that anyone who wants to have the Earth move along the circumference of a circle must first prove that it is not at the center of that circle. The next thing is for us to see whether the earth is or is not at that center around which I say it turns, and in which you say it is situated. And prior to this, it is necessary that we declare ourselves as to whether or not you and I have the same concept of this center. Therefore tell me what and where this center is that you mean.*

Often painful, albeit amusing, Salviati faces extraordinary resistance. If Salviati is correct, it is not merely a matter of redrawing a few celestial atlases. The implication is the abandonment of scholastic authority (Heilbron 2010). It is not easy to relinquish a paradigm.

A similar argument could be made for two contemporary world systems: **A** of symmetry and law; and **B** of information and adaptation. For a wide range of phenomena, it has not always seemed obvious which world system applies, and various attempts have been made to squeeze biology, medicine, economics, sociology, and political economy into **A**. Only in recent decades have these migrated towards a more natural home in **B**.

Reports from Two Worlds

Based on profound ideas from condensed matter, in conjunction with advances in the design of semiconductor fabrication plants, engineers are capable of building and controlling three nanometer (3nm) integrated circuits. This miniaturization exploits the evolution of complementary metal–oxide–semiconductors (CMOS), roughly following Moore's Law, leading to miniaturized transistor gate lengths of 3nm (around 250 million transistors per square millimeter) and below. Miniaturization has been accompanied by energy savings (lower supply voltages) and faster switching speeds (Jin 2023).

Viruses span a comparable range of scales from 20 to 200nm in diameter. This range accommodates genomes from several thousand to millions of base pairs (Holland 1998). The coronaviruses range from 50nm to 140nm (Holmes 1999). COVID-19 is about 100nm, the size of a transistor gate from 2000. COVID-19 encodes around fifty distinct proteins, each of which performs a rather specific function in the completion of the virus life cycle. The logic of gene regulation in COVID-19 does not come close to the diversity of operations that can be performed by a modern, vastly smaller, *integrated circuit* (**IC**).

How might we explain the discrepancy of control and understanding we possess over these two machines of comparable scale? It is not a question of numbers of components, or their complications and interactions, total functional repertoire, inorganic chemistry or biochemistry, or energetics. Something else is going on that makes us masters of miniature artifacts and victims of miniature organisms.

The key to understanding a virus is that it has evolved in order to achieve the unrelenting objective of replication in a variety of host cells. The virus is an agent with an evolution-ordained function. All practical differences between minerals, machines, and microbes follows from this fact of natural history.

A virus, a cell, an organism, an ecosystem, and a society all evolve—both organically and culturally—and exploit the out-of-equilibrium affordances of active matter. Because they evolve, they use sources of metabolic free energy efficiently to adapt. Through adaptation, the basic rules describing their properties and behavior are constantly changing, and doing so in response to stochastic processes and parameters in the environment. Unlike any purely physico-chemical system, the logic of their operation changes, and in such a way as to encode and anticipate salient features of the volatile world around them. There are laws of physics and there are rules of life—and rules are meant to be broken.

A virus is not at all like a modern integrated circuit; it is, rather, like an old integrated circuit in an organic computer (cell and organism), surrounded by a large team of programmers, all of whom respond to signals in the world (natural selection) so as to maximize viral growth rates. And the same basic logic can be applied to many complex systems.

The Second World System

Over the last century countless researchers have sought to capture the essence of complexity in ideas as wide-ranging as "self-organizing systems," "voluntary activity," "cybernetic control," "goal-seeking," "self-reproducing," "representational," "schematic," "autopoietic," "cognitive," and "information gathering and utilizing system" (IGUS).

See table 2 on pages 75–77.

Each of these ideas, and many more like them, attempts to identify the key character of mechanisms that promote persistent information-rich couplings, or long histories of strategic interaction. All slowly build upon simple out-of-equilibrium structures observed in nonlinear regimes.

Complexity science is also concerned with how life can be different from physics and chemistry but dependent on them: how to move **A** into **B** (**A→B**). One might just as well say how society comes to be different from brains and minds but remains dependent on them. The complex domain encompasses the world of self-organizing and evolving agents at all scales, from organisms to whole societies, and, increasingly, software and machines.

Complexity science is one of the most radical new scientific paradigms of the twentieth century. It includes paradoxes and challenges as puzzling as those in quantum mechanics and general relativity. Many of these puzzles are connected to the idea of emergence, including the origin of life from small molecule abiotic chemistry (the dynamics of **A→B**); how organisms coordinate into functional collectives; how "free will" (or the illusion thereof) might emerge from the biochemistry of brains; and how societies and their laws emerge from collectives of semi-autonomous agents.

Beyond these fundamental scientific difficulties there are enormous practical challenges. The application of complexity scholarship to global commons problems, including disease, climate, conflict, and political economy, is likely to be an essential component in any effort to ensure the prosperity and survival of life on Earth (Levin 1999; West 2018).

On the Origin of Paradigms

Where do new fields or disciplines come from? How do we establish when a series of inventions and discoveries warrants an entirely new paradigm, or whether existing ones might be modified to accommodate novelty? For example, at what limit of observation was it determined that physics was insufficient and that we needed to maintain the separate disciplines of chemistry and biology? Or why are there English, Italian, and other language departments and not just one monolithic linguistics department? How do we determine the variety and resolution of our epistemological commitments?

We should like to know what contributed to the origin and introgression of complexity science over the last century. And what new ideas, or new combinations of ideas, beyond those of physics, chemistry, economics, and evolution needed to be introduced and thereafter reconciled? Three ideas from philosophy help to flesh out the idea of a paradigm: language games, hermeneutic circles, and disciplinary matrices.

The philosopher Ludwig Wittgenstein suggested that these kinds of questions might be answered through the identification of new rule systems, or *language games*, as he called them (Ahmed 2010). A new language game is made up from a set of rules that are substantially different from, perhaps even mutually unintelligible to, those that came before. Different games have different features, and, as a practical matter, it is not always easy to determine by what sequence of behaviors we even identify the transition to a new game.

> *Compare chess with noughts and crosses. Or is there always winning and losing, or competition between players? Think of patience. In ball games there is winning and losing; but when a child throws his ball at the wall and catches it again, this feature has disappeared. Look at the parts played by skill and luck; and at the difference between*

> *skill in chess and skill in tennis. Think now of games like ring-a-ring-a-roses; here is the element of amusement, but how many other characteristic features have disappeared! And we can go through the many, many other groups of games in the same way; can see how similarities crop up and disappear* (Wittgenstein 1953, §66).

Scientific research is not merely a process of fitting new data into old models or into constellations of models comprising larger theories. New data often break the old models and theories. One cannot use the rules of chess to analyze a game of Go, nor can we use particle physics to understand genetics. Wittgenstein thought of games as the supreme metaphor for the specificity and context-dependence of formal thought: new possibilities need to be explained with new rules. And, we might add, in what "language" are these new rules written—natural language, mathematics, computer code?

In 1900 Wilhelm Dilthey described the experience of understanding a new phenomenon through a strict context-dependence in terms of a *hermeneutic circle* (Palmer 1969):

> *It is at this point that the central difficulty of all exegetical practice makes itself felt. The whole of a work is to be understood from the individual words and their connections with each other, and yet the full comprehension of the individual parts presupposes comprehension of the whole* (Dilthey and Jameson 1972, 243).

Dilthey was in effect pointing out a "chicken-and-egg" problem for understanding: parts are required to comprehend the whole but the whole is needed to make sense of the parts. This seems to be particularly apropos of complexity, which, by some definitions, is the science associated with the maxim "the whole is greater than the sum of its parts."

Thomas Kuhn (2012) expanded on Dilthey's hermeneutics and Wittgenstein's language games to explain how scientific ideas evolve through the persistence and overthrow of *paradigms* constituted by a *disciplinary matrix* describing how ideas relate to each other. The matrix describes how ideas are mutually dependent, and how some might be reduced into elementary facts and others aggregated into synthetic propositions. It is a defining feature of paradigms (like language games and herme- neutic circles) that they are incommensurable, or discordant, with one another.

> *For present purposes I suggest "disciplinary matrix": "disciplinary" because it refers to the common possession of the practitioners of a particular discipline; "matrix" because it is composed of ordered elements of various sorts, each requiring further specification. (*Kuhn 2012, 181)

> *Though the strength of group commitment varies, with non-trivial consequences, along the spectrum from heu- ristic to ontological models, all models have similar functions. Among other things they supply the group with preferred or permissible analogies and metaphors. By doing so they help to determine what will be accepted as an explanation and as a puzzle–solution; conversely, they assist in the determination of the roster of unsolved puzzles and in the evaluation of the importance of each* (Kuhn 2012, 183).

Disciplinary matrices might be likened to a mechanical diagram of an automobile illustrating how each part is connected, assembled, and made interoperable with other parts. Quantum mechanics is a paradigm, plate tectonics is a paradigm, organic chemistry is a paradigm, and neoclassical economics is a paradigm. None of these fields is defined by a single anomalous experiment, model, or idea, but by a matrix of compatible elements. When an incom- patible observation or model is thrown into the mix, it threatens to demolish the matrix. This more often than not leads to the rejection of the novel element, or what the philosopher of science Gunther Stent called *prematurity*.[2]

Whether an idea is accepted or rejected depends to a large extent on how central it is to keeping a paradigm connected. A car might function without its heater, but it is worthless without an engine block. Adding a large battery to a combustion engine does not make sense.

And yet multiple paradigms can coexist as long as each provides instrumental value in its respective domain. We launch satellites with classical mechanics, navigate by satellite using relativistic mechanics, and exploit understanding of quantum mechanics to build semiconductor-based photovoltaics powering the satellite. Three paradigms encased in one celestial artifact. Incommensurability need not imply incompatibility. This is one of the explanations for pluralism in science: different paradigms play different roles and their practical value often outweighs their incommensurability.

Periods of "normal" science consist in modifying or adding to existing paradigms—adding rows and columns to the matrix. Periods of "revolutionary" science consist in rewiring concepts and forming novel communities, through this process generating, in Kuhn's language, a "plurality of worlds."

From the perspective of rule systems and paradigms, disciplines like geology and anthropology are not simply two different views onto reality using the same underlying universal rationality. They are describing two different, albeit overlapping, sets of empirical principles, including different tools for investigating emergent worlds. Chemistry is not just physics-at-scale,

---

[2] "A discovery is premature if its implications cannot be connected by a series of simple logical steps to contemporary canonical [or generally accepted] knowledge." —Gunther Stent, *Prematurity in Scientific Discovery: On Resistance and Neglect*

but a set of mechanisms that requires distinct and extra-physical principles and models to be usable. Similarly, biology is not just chemistry-at-scale, and so forth through the spatial and temporal hierarchy.

The history of complexity science represents a revolutionary transformation in our way of understanding the world that forged four areas of research into a new way of seeing the world: a new world system. We might abbreviate these as: evolution, entropy, dynamics, and computation. In the process of connecting these areas, principles from each had to be reconciled, modified, and extended. This resulted in a greater understanding of the concenter of complex systems—*the theory of far-from-equilibrium, purposeful machines.*

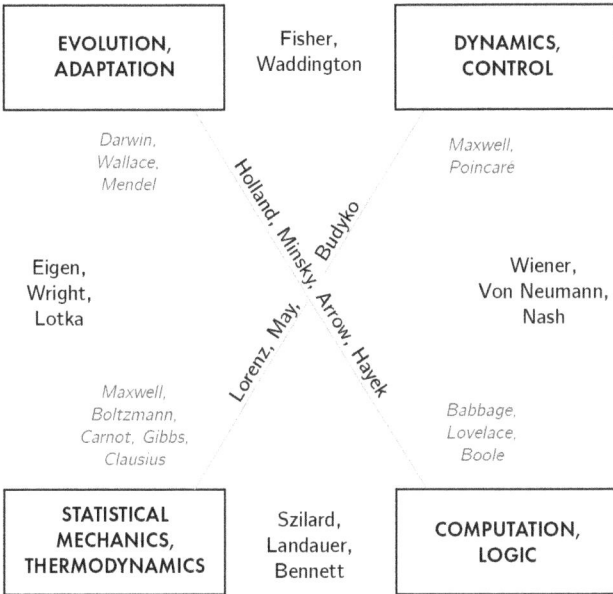

Figure 2. The four historical pillars of complexity science (Evolution and Adaptation; Dynamics and Control; Statistical Mechanics and Thermodynamics; Computation and Logic) as established in the nineteenth century and labeled with a few of their most consequential contributors (dashed lines). Complexity science emerges from the concerted effort to connect these four domains through the invention or union of principles. A few illustrative "intellectual edges" in the graph are shown where each is contributed by a foundational paper: Fisher and Waddington connecting dynamical systems with evolutionary theory; Eigen, Lotka, and Wright connecting statistical dynamics or thermodynamics to evolutionary theory; and so forth. Complexity science emerges historically as the spanning set defined by these principles.

Constructing the Complexity Paradigm

New paradigms can come into existence through a number of parallel processes, including the discovery or proposal of radically new phenomena (e.g., genetic dominance, quantum entanglement, localization of function in the brain), from the development of new methods and models for studying phenomena (e.g., population genetics, Bell's theorem, linguistics, connectionism), and from the unification of existing fields through the invention of laws or principles that reconfigure the boundaries of understanding (e.g. evolution, game theory)—rewiring the disciplinary matrix (see Kuhn 2000; Callebaut and Pinxten 2012; Brad Wray 2021; and Guerra, Capitelli, and Longo 2012 for review of this area).

Take molecular biology. Rosalind Franklin's use of X-ray crystallography to produce Photo 51 provided somewhat cryptic information about the structure of DNA (Pederson 2020). James Watson and Francis Crick used Franklin's insights as the basis for their inference of a double-helical molecule of inheritance. The structure of the molecule led more or less directly to a new mechanism of genetic replication. This mechanism could be made compatible with both meiosis and mitosis during cell division, and, more circuitously, it could provide a justification for the "central dogma" of molecular biology (the now-refuted one-directional flow of information from DNA to RNA to proteins) (Shapiro 2009). Hence a new technique revealed a new structure that in turn supported a novel suite of functions and thereupon a rather profound mechanistic refutation of Lamarckian inheritance (the genetic transmission of acquired characters).

The same logic could be unfolded for numerous fields, including Max Planck's study of black-body radiation, Vera Rubin's observations of galaxy rotation to infer the existence of dark matter, and Carl Woese's use of sequencing and phylogenetic techniques to identify a third branch of life using ribosomal RNAs.

For complexity science, the new phenomena relate principally to machines, organisms, and collectives with purpose: function or *teleology*.[3] The new methods investigate the laws and rules that govern these machines, including their efficiency, predictability, control, and coordination. And unification has taken the form of frameworks that balance and connect the ideas of evolution, entropy-production, nonlinear dynamics, and computation. These concepts are typically expressed in terms of processes of origination, optimization, stability, robustness, resilience, energy dissipation, regulatory mechanics, informational constraints, and computational costs and limits. These elements of the complexity matrix are familiar to most researchers in the field despite working on very different problems.

One of several public-facing "manifestos" for this new integrated science, an effort to describe this new matrix, was Norbert Wiener's 1948 *Cybernetics: Or Control and Communication in the Animal and the Machine.* Written in eight chapters, starting with classical mechanics, and proceeding through statistical mechanics, communication, control theory, computation, and ending with models of society. Another was Claude Shannon and Warren Weaver's 1949 *The Mathematical Theory of Communication,* written in two long sections, the first laying out the mathematics of information theory as applied to machines, and the second, limits and possible extensions into a theory of semantics and behavior in living systems.

Complexity science seeks to understand an adaptive ontology— infiltrated by far-from-equilibrium, nonlinear, dissipative mechanisms, found in both nonliving and living systems. In the Kuhnian sense, the complexity paradigm is incommensurable with many

---

[3] *Teleology* refers to a goal-oriented explanation, interpreting structures or dynamics in terms of their functions. In biological contexts, teleology is a consequence of natural selection, and in cultural contexts a result of learning and planning. Typically teleology implies internal states supporting conditional branching—pursuing different behaviors as a result of contingent history and context.

that came before and yet remains compatible with them. In order to address the limitations of existing ideas, models, and frameworks, new effective laws have needed to be discovered. These account for the emergent properties that Philip Anderson, in his paradigm-supporting paper "More Is Different" (1972) sought to describe. New effective theories might not be fully computable, and many will be "inelegant" by the standards of fundamental theory. One could describe a bacterium like *Escherichia coli* as a point mass, but it would be pointless. Much that is of interest beyond the physico-chemistry is related to the historical logic of adaptive mechanisms.

There has been need for new methods, system-level descriptions, and new technical insights to grapple with broken symmetry, non-ergodicity, dissipative dynamics, and their sequels. These are the crucial epistemological and auxiliary considerations attendant on building up the disciplinary matrix of complexity.

Reducing complexity science exclusively to methods diminishes its paradigmatic status. These have been techniques for studying out-of-equilibrium patterns or new quantitative techniques for finding statistical order in large datasets. At its worst, this tendency equates the many principles of complexity with the search for a universal metric that might apply to all observables: the complexity of a brain, societies, and technologies, all on one single axis of quantification. This is a case of techniques substituting for theories. It would be like trying to measure the degree of "chemistry" in all chemical reactions or the quotient of "anthropology" in a society.

There are schools of thought in which everything is mathematics, everything is physics, everything is poetry, and everything is religion. These kinds of framing of reality are lacking in richness, subtlety, modesty, and value. The last thing we want is for everything to be complexity. An important step towards this restraint is not to confuse the application of a method with a field of inquiry. The insights of Dilthey, Wittgenstein, Kuhn, and Anderson all help us to understand why this is the case. Complexity science should help us to understand why a plurality of paradigms is not only of utility but inevitable.

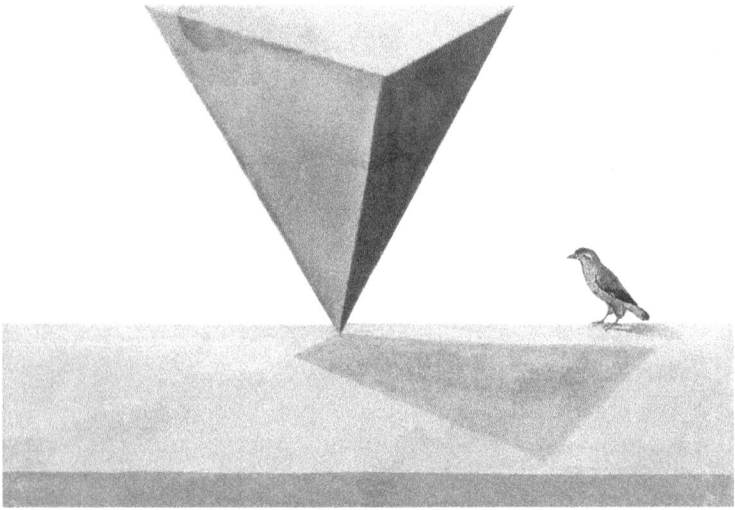

Prefoundation

Mechanical Matter

The development of complexity science was strongly influenced by the emerging industries, records of scientific exploration, and new theories produced in the age of steam and machines. If modern physics and chemistry have their roots in the scientific revolution of the seventeenth century, complexity science has its roots in the Industrial Revolution and its aftermath in the nineteenth century. This was the age of working matter: manufactured machines and evolved organisms. In the nineteenth century parallels were consciously sought that might connect manufactured technology with the mechanics of life, and entirely new vocabularies needed to be invented to describe this interface. Tom Peters (1996, 356), writing on the topic of machines and manufacturing in the nineteenth century, explains how these ideas were connected through technology:

> The machine, war, and life are concepts that we can trace posteriori in older building processes. But their conscious application to construction dates to that period in which machine-making, war, and biology themselves evolved from arts to technologies.

Charles Babbage (1832), a notable inventor and analyst of technology, who would exert a lasting influence on Karl Marx's theories of industry, wrote optimistically on the substitution of organic traits with mechanical prosthetics:

> The hand of man is now too slow for the demands of his curiosity, but the power of steam comes to his assistance.
> (Babbage 1832, 218)

The machine was self-evidently a purely physical system, entirely constructed, and yet described in a language different from physics. The requirements of mechanical work engendered a new vocabulary of tools and equipment emphasizing concepts particular to these artifacts. These include the ideas of cost, production, construction, efficiency, control, utility, durability, life-span, value, and market. Latently, each of these concepts rooted in technology would provide the seeds for the development of entirely new natural scientific and mathematical theories and models.

Milestones in the Prefoundations of Complexity

Searching for prequels to complexity science in the nineteenth century—the prefoundational period—we detect the strongest signals in four research areas. These include ideas relating to entropy by Sadi Carnot, Rudolf Clausius, James Clerk Maxwell, Josiah Willard Gibbs, and Ludwig Boltzmann; theories of evolution and adaptation in the work of Charles Darwin, Alfred Russel Wallace, and Gregor Mendel; the mathematics of nonlinear dynamics and control in the work of Maxwell and Henri Poincaré; and the concept and machinery of difference and analytical engines and logic in Charles Babbage, Ada Lovelace, and George Boole.

All fields are porous, and several key figures from the eighteenth century—most notably Gottfried Leibniz (monadology and mechanical calculators), Alexander von Humboldt (biogeography and ecology), Thomas Malthus (population growth and collapse), and Adam Smith (invisible hand and division of labor)—might justly be teleported into the nineteenth century for purposes of extending the primary contributors to the new science; but the formal, or mathematical, exposition of key ideas is very much of the nineteenth century.

From a contemporary, disciplinary perspective, we might group these four areas into the separate histories of physics, chemistry, engineering, biology, mathematics, and computation. It is not clear that this is what these pioneers would have wanted. After all, Darwin and Wallace thought of selection in terms of dynamical regulation and physical laws, Babbage and Boole thought about natural law and mind in terms of calculation and dynamics, Gibbs thought of statistical physics in terms of probability theory, Boltzmann thought of evolution in terms of statistical physics, and along with Poincaré, interpreted mathematics in light of evolution. Nevertheless, complexity science proper is a twentieth-century program in the unification of natural and engineering sciences bearing directly on out-of-equilibrium and living phenomena, including a variety of extensions into society, technology, and history.

Four pillars—*entropy* (**E**), *evolution* (**V**), *dynamics* (**D**), and *computation* (**C**)—had been established in the nineteenth century, and much twentieth-century effort consists in integrating these four areas of research: (1) the principles of order and disorder in relation to the production of mechanical work; (2) the mathematics of the evolutionary and adaptive process; (3) the regulation, control and prediction of nonlinear dynamical systems; and (4) the informational and computational requirements of purposeful or intelligent behavior.

The task has been not only to integrate these four, but, where possible, to discover general principles that might subsume them. These principles include understanding, stability, order, predictability, adaptability, function, aging, and collapse in far-from-equilibrium systems and "teleonomic" matter.

In the following sections, I briefly enumerate the contributions of each of these researchers with an emphasis on ideas that have proven to be of significance in twentieth-century complexity science.

CARNOT, BOLTZMANN, MAXWELL, CLAUSIUS, AND GIBBS: THE EFFICIENCY OF MACHINES AND THE ARROW OF TIME

1824

Sadi Carnot, an engineer, concerned himself with the problem of how to maximize the amount of mechanical work performed during the transfer of heat from a boiler to a condenser in a steam engine. The problem was highly practical but his insights enormously general. Carnot intuited much that would come later with the development of the entropy concept. Carnot verbally introduced the idea of a reversible thermodynamic cycle which we now think about in terms of two isothermal processes (constant temperature) and two adiabatic processes (constant heat or entropy). Clapeyron translated Carnot's work into mathematics and visualized the cycle using the Watt Indicator diagram. Carnot understood that only the differences in temperature between the two reservoirs mattered, and not the medium (gas, fluid, etc.) performing the work. Carnot showed that the key to maximizing the efficiency of any work cycle is to minimize the loss of heat (in modern terms strive to keep the entropy of the system constant over the cycle). The Carnot cycle is an ideal engine and can only be approximated in practice. The discrepancy between the ideal cycle and physical reality provides an instrumental definition of the second law of thermodynamics.

1871

James Clerk Maxwell in his *The Mechanical Theory of Heat* set out to establish the status of thermodynamics as a physical theory. In practice this meant "investigating the relations between the thermal and mechanical properties of substances" or exploring the epistemological limits of grounding thermodynamics in classical mechanics. In order to accomplish this Maxwell adopted his method of "physical analogies" whereby

Table 1. A chronology of some of the major figures and milestones from the prefoundational period. This table is far from comprehensive yet represents a fairly uniform sample of ideas of incontestable importance in seeking to understand the roots of complexity. This is a necessary, albeit insufficient, nineteenth-century proto-complexity reading list.

YEAR | AUTHORS	CODE LETTERS | CORE COMPLEXITY CONNECTIONS	KEY IDEAS IN PAPER
1824 Sadi Carnot	E | Thermodynamics; statistical mechanics	The efficiency of heat engines; the Carnot cycle
1837 Charles Babbage	C | Computation; cognition	Analytical engine; mechanical general-purpose computer
1841 Charles Babbage	C | Computation; cognition	The Ninth Bridgewater Treatise: on mathematical induction; low probability events; and miracles (simulation theory)
1843 Ada Lovelace, Luigi Menabrea	C | Computation; cognition	Analytical engine suggests a generalized science of "operations" including mechanical creativity (in music)
1847 George Boole	C | Computation; cognition	Theory of probability; algebraic methods of logic
1854 George Boole	C | Computation; cognition	The fundamental operations of mind as described using a probabilistic and universal logic; thought as a natural law
1858 Charles Darwin, Alfred Wallace	V | Evolution; adaptation	The principle of natural selection; selection as analogous to a centrifugal governor

YEAR | AUTHORS	CODE LETTERS | CORE COMPLEXITY CONNECTIONS	KEY IDEAS IN PAPER
1859 Charles Darwin	V | Evolution; adaptation	On the origin of species; selection as analogous to a physical force (gravity)
1865/6 Gregor Mendel	V | Evolution; adaptation	The inheritance/trans-mission and discrete segregation of heritable factors
1868 J. Clerk Maxwell	E | Dynamical sys-tems; stability and chaos	The stability of integral feedback control; nega-tive feedback
1871 J. Clerk Maxwell	E | Thermody-namics; statistical mechanics	The mechanical basis of statistical phenomena; "Maxwell's demon"
1872 Ludwig Boltzmann	E | Thermody-namics; statistical mechanics	The increase in entropy of a system towards the Maxwell–Boltzmann distribution (H-theorem)
1875 J. Willard Gibbs	E | Thermody-namics; statistical mechanics	The purely probabilistic interpretation of the second law
1879 Rudolf Clausius	E | Thermody-namics; statistical mechanics	The mechanical theory of heat, the theorem of equivalence of transfor-mations
1890 Henri Poincaré | D	Dynamical systems; stability and chaos	The three-body problem; transverse homoclinic orbits; Poincaré maps

all hypotheses of assumed collective molecular geometry and motion were grounded in the constraints of Newton's laws and the conservation of energy. By pursuing an hypothesis involving many spherical particles engaged in perfectly elastic collisions, Maxwell was able to deduce the stationary distribution of particle velocities (the distribution left unchanged by further collisions). Both Clausius and Boltzmann had sought to use this kind of reasoning to explain the law-like status of Carnot's observation that heat cannot pass from a colder body to a warmer without some other change occurring. They were unsuccessful. Maxwell explained why this was so through his thought experiment with a small mechanism (at the same scale as the particles) able to distinguish each particle and guide and control their actions (never violating the conservation of energy) so as to violate the second law: allowing heat to pass from a colder (low-energy) state to a warmer (high-energy) state. The importance of what Lord Kelvin called "Maxwell's demon" was to show that thermodynamics had to be built on statistical foundations, to be a highly probable process, and to give up its ambition at inter-theoretic deduction from more fundamental principles. In this regard, Maxwell's work foreshadowed one of the core ideas behind emergence, as described in Philip Anderson's paper "More Is Different" a century later in 1972. It also foresaw the possibility and potential challenges of building mechanisms to capture "thermal information" in nonequilibrium computing devices.

1872

Boltzmann was placed in a quandary through the insights of Maxwell. He had made it his objective to reconcile time-reversible mechanics of point masses with thermodynamics without introducing microscopic stochasticity. On the other hand, Boltzmann needed to start with probability in order to derive his expression for entropy increase: the *H-theorem*.[4] Boltzmann

---

[4] The H-theorem was developed by Ludwig Boltzmann to describe the evolving energy distribution of molecules. Over the course of numerous elastic collisions with associated transfers of energy, the speed distribution of molecules converges on a stationary distribution—the Maxwell–Boltzmann distribution—which minimizes the value of H (increases the entropy).

developed his ergodic conjecture as an interface between these levels of descriptions: leave particles bouncing around for long enough and through "molecular chaos" they will even out in space. Now replace the microscopic description with a distributional (probabilistic) description and operate at the probabilistic level. Note that classical mechanics is abandoned in this move towards statistics. Whereas Maxwell used the "demon" to demonstrate the intrinsically statistical nature of the second law, Boltzmann outlawed demons by replacing microscopic observables with continuous density functions. Boltzmann was thereby free to write down the dynamics of a gas through a time-dependent density of particles in an appropriate phase space (positions and velocities), and show that H (the distance to the Maxwell distribution) always decreases in time, or the entropy (-H) increases in time. Thus Boltzmann sought to derive a time-asymmetric irreversible process from an underlying time-symmetric reversible mechanics. Regardless of the degree of success of his project, the profound implication of Boltzmann's work was to explain the apparent asymmetry of the past and the future and thus lay the foundations for some of the most important concepts in complex systems: causality, inefficiency, robustness, resilience, aging, and death.

1875

Josiah Willard Gibbs did not share the Maxwell–Clausius–Boltzmann preoccupation with Newtonian fundamentals. Reviewing his ideas in the 1902 monograph *Elementary Principles in Statistical Mechanics,* Gibbs jumped straight into probabilities:

> *For some purposes, however, it is desirable to take a broader view of the subject. We may imagine a great number of systems of the same nature, but differing in the configurations and velocities which they have at a given instant, and differing not merely infinitesimally,*

> *but it may be so as to embrace every conceivable combi-*
> *nation of configuration and velocities. And here we may*
> *set the problem, not to follow a particular system through*
> *its succession of configurations, but to determine how the*
> *whole number of systems will be distributed among the*
> *various conceivable configurations and velocities at any*
> *required time, when the distribution has been given for*
> *some one time. . . . Such inquiries have been called by*
> *Maxwell statistical.* (Gibbs 1902, vii-viii)

Freed in some sense from traditional "physics," Gibbs could also free himself from the coils of historical thermodynamics, and in so doing move considerably beyond it. Throughout the 1870s, Gibbs introduced a series of new formalisms and ideas, including the vector calculus, the "thermodynamic equation of a fluid," the "surface of absolute stability," the "Gibbs paradox," and "Gibbs free energy" (Gibbs 1873, 1875; Wilson and Gibbs 1901). In many ways Gibbs is the first post-emergence thinker—having no need to pursue reductionist physics in order to discover a new fundamental physics, expressed through a purely effective (non-fundamental) theory.

1879

The career of Clausius spans the contributions of Carnot, Maxwell, Boltzmann, and Gibbs. The name entropy was provided by Clausius in 1865 as well as many of its practical implications. Clausius concluded his entropy paper with the famous couplet: "The energy of the universe is constant. The entropy of the universe tends to a maximum." Herein lies one of the many fractures dividing the nineteenth century from the twentieth century: we might call it the transition from simple symmetry to complex broken symmetry. *The Mechanical Theory of Heat* from 1871 represents a mature form of mechanical thermodynamics with a strong emphasis on work and might be read as the culmination of the engineering concerns of Carnot. It is full of engines, heat baths, and tubes. One idea that Clausius championed, and that has faded from discussion, is his idea of "disgregation," a concept describing the geometric dispersion of molecules in the generation of mechanical work. Clausius saw entropy as a sum of a heat term and a "disgregation" term, and felt that this term captured something more fundamental about the hidden mechanics of molecules than their energy.

BABBAGE, LOVELACE, AND BOOLE: CALCULATORS, COMPUTERS, AND THE LOGIC OF MIND

1837–1851

Charles Babbage, having made a name for himself with the publication in 1824 of *On the Economy of Machinery and Manufacture*, becoming the Lucasian Chair of Mathematics at Cambridge in 1828, and co-founder of the Royal Astronomical Society, begins his contribution to computation with plans for a *Difference Engine* (**DE**). A device to be used in the automatic calculation of the Nautical Almanac, comprising tables of data bearing on the determination of longitude while at sea. After studying Joseph Marie Jacquard's 1805 punch-card mechanism for weaving arbitrary designs when attached to looms, Babbage conceived of his dual store–mill mechanism for a general-purpose computer, the *Analytical Engine* (**AE**). Ada Lovelace, having translated Luigi Menabrea's lucid description of Babbage's AE from a series of 1840 lectures in Turin, wrote, "We may say most aptly that the Analytical Engine *weaves algebraical patterns* just as the Jacquard-loom weaves flowers and leaves. Here, it seems to us, resides much more of originality than the Difference Engine can be fairly entitled to claim" (Taylor 1843, 696). Lovelace was likely the first to conceive of the art of computer programming, describing an algorithm for computing the Bernoulli numbers, as well as speculating on the computational generation of musical and art works; a computational perspective going far beyond look-up tables and ephemerides. Babbage's sweeping vision of computation led him to write *The Ninth Bridgewater Treatise*, an attempt to demonstrate that "there exists no such fatal collision between the words of Scripture and the facts of nature." The monograph is the first thorough working out of what we now call *simulation theory*. It presents a computational analog to Maxwell's demon in the form of programs that conditionally branch to produce rare events—miracles—based on purely mechanical principles. The profundity of Babbage's work remained in dormancy until the 1940s and 1950s when they were rediscovered by Howard Aiken and Alan Turing.

1847–1854

George Boole's professional career represents an approach to mathematics and logic distinct from those of his contemporaries. For Boole, Professor of Mathematics at Queen's College in Cork from 1849, mathematical research represented a means of investigating "the fundamental laws of those operations of the mind, by which reasoning is performed." This was the stated aim of his 1854 monograph, *An Investigation of the Laws of Thought*; in it he constructed a framework in which "processes of symbolical reasoning are independent of the conditions of their interpretation." Boole's approach to thought closely resembles Claude Shannon's later approach to both circuit design and information—both sought to make progress by eliminating semantic content from automatic procedure, and constructing these procedures independently from machinery. Boole's earlier work, *The Mathematical Analysis of Logic* (1847), starts by making the same observation: "Thus the abstractions of the modern Analysis, not less than the ostensive diagrams of the ancient Geometry, have encouraged the notion, that Mathematics are essentially, as well as actually, the Science of Magnitude." Boole's technique was to introduce a new algebra (related to Leibniz's algebra of concepts), which we now describe in terms of a binary-valued Boolean logic, or an algebra of sets. In concluding his work, Boole alluded in embryo to the topic of mathematical intuition in connection to the power of "unconscious" thought—a topic in torpor for a century until the publication of *The Psychology of Invention in the Mathematical Field* by Jacques Hadamard in 1945.[5] Boole wrote, "It is not contended that it is necessary for us to acquaint ourselves with those laws in order to think coherently, or, in the ordinary sense of the terms, to reason well. Men draw inferences without any consciousness of those elements upon which the entire procedure depends" (Boole 1854, 422–423).

---

[5] Logic and the unconscious: Hadamard's interest in the role of the unconscious in analytical thought can be traced back to the writings of Arthur Schopenhauer in his essay, "Transcendent Speculation on the Apparent Deliberateness in the Fate of the Individual" (1851), further developed by Eduard von Hartmann in "Philosophy of the Unconscious: Speculative Results According to the Induction Method of the Physical Sciences" (1869) and applied to mathematics by Henri Poincaré in his book *Science and Method* (1908) in a chapter on mathematical creation.

DARWIN, WALLACE AND MENDEL: EVOLUTION, ADAPTATION, AND GENETIC TRANSMISSION

1858–1859

Charles Darwin and Alfred Russel Wallace introduced the essential elements of their theory of natural selection, competition, divergence, speciation, and sexual selection in 1858. The theory placed material laws at the center of biology as they had been in physics and chemistry for over a century. Natural history was thereby reconciled with a universal mechanism. Darwin and Wallace's view of selection is as a powerful distributed machine for making discriminations: "Now suppose there were a being who did not judge by mere external appearances, but who could study the whole internal organization, who was never capricious, and should go on selecting for one object during millions of generations" (Darwin and Wallace 1858, 51). The selective environment in this account is not reducible to a simple phenomenological variable—the Malthusian parameter—but is conceived of as a high-dimensional filter capable of detecting countless internal degrees of freedom. Darwin described this environment in terms of "an entangled bank, clothed with many plants of many kinds . . . elaborately constructed forms, so different from each other, and dependent upon each other in so complex a manner, have all been produced by laws acting around us" (Darwin 1859, 489). Darwin was describing his own "demon" in analogy to Maxwell's Newtonian demon, capable of seeing into the "black box" of the organism. The way that selection achieves this X-ray-like capability is in terms that what we would now think of as the *ergodic hypothesis*, whereby countless generations—time—allow that the space of microstates ("the whole internal organization") is explored and evaluated. The mutation-selection process is an algorithmic procedure for generating and propagating neutral families (ensembles) of approximate solutions to adaptive challenges in high-dimensional fitness landscapes, a perspective subsequently codified in modern approaches to genetic algorithms and genetic programming.

1865–1866

Gregor Mendel's paper of 1866 provided the "genetic" basis for heritability that enabled the Darwin–Wallace mechanism to function (Mendel 1866). Darwin's own theory of pangenesis proved to be incorrect. Mendel's mechanism was very popular for espousing an idea resembling an atomic theory of variation. The implications of Mendel's theory are captured by a simple recurrence equation whose fixed point has come to be known as the Hardy–Weinberg equilibrium. It is a useful null expectation under very strict conditions of invariance. Mendelian genetics is, however, highly non-ergodic, or saltationist, and explores a very small volume of sequence space. In the early twentieth century, Raphael Weldon suggested a more realistic theory, incorporating multiple internal and external causal factors, allowing for approximately continuous variation in organisms. Weldon died before his book, *Theory of Inheritance,* was published. Structurally similar modifications were made later by such figures as C.H. Waddington and James Mark Baldwin, key contributors to complexity science (Radick 2023).

MAXWELL AND POINCARÉ:
THE STABILITY OF MACHINES AND INSTABILITY OF
SOLAR SYSTEMS

1868

James Maxwell's paper "On Governors" represents the first rigorous analysis of feedback control, the founding paper of what later would be called cybernetics, and a pioneering application of dynamical systems to modeling machines—as it happens, the same machines that inspired Sadi Carnot to invent the field of thermodynamics: steam engines. Maxwell's interest was piqued by the challenge of maintaining a constant speed of a steam engine in the face of uneven surfaces and supplies of fuel. Maxwell used differential equations to describe two forms of feedback mechanism controlling the delivery of fuel to an engine: moderators and regulators. He was able to show, by analysis of the characteristic equation of these systems, that differential control (widely adopted) was unstable, whereas integral control was stable. Maxwell's contribution was ignored for over eighty years until it was rediscovered by Norbert Wiener in 1948. Maxwell's work introduced several concepts that lie at the heart of the study of complex systems: (1) an elementary form of agency whereby a purely physical system pursues a target by minimizing an error function; (2) an emphasis on stability which undergirds ideas of robustness, error-correction, and resilience; and (3) an insight into the role of a system memory (integral control) able to store a cumulative history of error in order to regulate a system reliably.

1890

The question of stability also motivated Henri Poincaré's analysis of the three-body problem. If we assume three gravitationally interacting bodies, all subject to Newton's laws of motion and gravitation, of arbitrary mass and initial condition, can we write down a solution for the three orbits into the past and into the future? The answer discovered by Poincaré remains the answer today: no. The equations of motion are nonintegrable and no general closed-form solution exists. However, in studying this system and simplified subproblems, Poincaré was the first to conceive of the idea of chaos, recognizing the three-body problem's extreme sensitivity to initial conditions—related to problems of diverging trajectories—compromising the ability to mathematically integrate out future orbits. The larger philosophical implications of Poincaré's findings relate to prediction and, by extension, free will. As Poincaré wrote himself in 1903: "If we knew exactly the laws of nature and the situation of the universe at the initial moment, we could predict exactly the situation of that same universe at a succeeding moment . . . it may happen that small differences in the initial conditions produce very great ones in the final phenomena. A small error in the former will produce an enormous error in the latter. Prediction becomes impossible . . ." (Poincaré 1903, 68). The development of nonlinear dynamics subsequent to Poincaré, including the crucial contributions of Edward Lorenz in the 1960s and Robert May and Mitchell Feigenbaum in the 1970s, established the generality of chaos. And the very high dimensionality of most complex systems, including the diversity of interactions among their parts, describes worlds where the elegant mathematics of celestial mechanics hits a wall. Perhaps descends into an analytical singularity?

CONNECTIONS AND SOCIAL NETWORKS

One of the appealing features of nineteenth-century science is the way that its modest scale promoted familiarity outside of the vehicle of the journal system. Many of the major contributors in table 1 (see pp. 36–7) knew each other personally, if not in person, then at a short remove, through universities, institutions, and scholarly societies. Hence ideas and influence flowed as an advance wave through social and professional networks before publications became generally available.

Charles Babbage was one of three founders of the Cambridge Analytical Society (along with William Herschel and George Peacock) bent on a mission to bring the European emphasis on algebraic abstraction and analysis to a Britain hitherto dominated by physical intuition and synthetic geometric reasoning. Babbage (1864), like many other students of mathematics at the time, had hoped that Cambridge would open his mind to the full range of research on the Continent but often found its materials rather provincial:

> *Thus it happened that when I went to Cambridge I could work out such questions as the very moderate amount of mathematics which I then possessed admitted, with equal facility, in the dots of Newton, the d's of Leibnitz, or the dashes of Lagrange. I had, however, met with many difficulties, and looked forward with intense delight to the certainty of having them all removed on my arrival at Cambridge. [. . .] I had heard of the great work of Lacroix, on the "Differential and Integral Calculus," which I longed to possess . . . .* (Babbage 1864, 26)

Through the formation and meetings of the Analytical Society, its members would exert a slow but sustained pressure, largely through the Royal Society in London, on British mathematical thought and notational conventions. An obvious example is the adoption of the differential form of Maxwell's equations (1865), which makes use of the very convenient Laplace operator to express his theory of electromagnetism. And Maxwell continued to use the differential form in his more popular books, including *An Elementary Treatise on Electricity* (1881). Similarly, over the course of his career, George Boole would feature the differential notation for the calculus, which he used analogically to introduce his theory of elective symbols and functions, which one might broadly interpret as binary values and Boolean operators.

Charles Babbage was also a notorious raconteur, bon vivant, and London socialite, throwing regular weekend parties at 1 Dorset Street. Guests included Ada Lovelace, Michael Faraday, Charles Lyell, Charles Dickens, Felix Mendelssohn, Charles Darwin, and literally hundreds more (Snyder 2011). These events featured both scientific discussions and free-ranging polemics:

> *I used to call pretty often on Babbage & regularly attended his famous evening parties. He was always worth listening to, but he was a disappointed & discontented man; & his expression was often or generally morose. I do not believe that he was half as sullen as he pretended to be. One day he told me that he had invented a plan by which all fires could be effectively stopped, but added, — "I shan't publish it — damn them all, let all their houses be burnt." They all were the inhabitants of London.* (Charles Darwin 1876, 133)

Researchers in this period would also meet at Exhibitions. Boole and Babbage met at the 1862 International Exhibition in London, where Babbage displayed diagrams and parts of his unbuilt Analytical Engine:

My dear Sir, It is a source of regret to me that I was quite unable to avail myself of your kind invitation to call upon you on my return from Cambridge to London … Meanwhile, I shall endeavor to acquaint myself with Menabrea's paper and the principle of the Jacquard loom. But I cannot allow this opportunity of writing to you to pass without thanking you very warmly for the kind explanations you gave me of the working of the Difference Engine, and without saying that it was a pleasure and an honour to me to meet you (Boole 1862).

Boole's books on the calculus of logic and thought were written in 1847 and 1854 and could not therefore be inspired by meetings with Babbage. Boole's critical predecessors, according to his own writing, were Aristotle and Augustus De Morgan, neither of whom in their logical publications were much concerned with physical reality. Nevertheless, *An Investigation of the Laws of Thought* from 1854 did purport to be a treatise on the natural intellect and not merely an exposition on a new mathematical logic. Boole does not at any point mention physics, chemistry, or natural history, or any of the scientist practitioners he would have met in society or their opinions on his ideas. The only mention of empiricism, in an age dominated by experiments, comes in a general allusion to experimental evidence in support of scientific knowledge:

Thus the necessity of an experimental basis for all positive knowledge, viewed in connexion with the existence and the peculiar character of that system of mental laws, and principles, and operations, to which attention has been directed, tends to throw light upon some important questions by which the world of speculative thought is still in a great measure divided. How, from the particular facts which experience presents, do we arrive at the general propositions of science? (Boole 1854, 402)

It is interesting to note Alfred Russel Wallace's comments on mathematics. They provide a clue to why Boole ignored underlying physiological and behavioral mechanisms and dispositions:

> *We have to ask, therefore, what relation the successive stages of improvement of the mathematical faculty had to the life or death of its possessors; to the struggles of tribe with tribe, or nation with nation; or to the ultimate survival of one race and the extinction of another. If it cannot possibly have had any such effects, then it cannot have been produced by natural selection . . .*

> [...]

> *We conclude, then, that the present gigantic development of the mathematical faculty is wholly unexplained by the theory of natural selection, and must be due to some altogether distinct cause.* (Wallace 1889, 466–467)

Wallace concludes his pro-Darwinian monograph with an anti-Darwinian appeal to extra-physical capability:

> *These three distinct stages of progress from the inorganic world of matter and motion up to man, point clearly to an unseen universe—to a world of spirit, to which the world of matter is altogether subordinate. To this spiritual world we may refer the marvelously complex forces which we know as gravitation, cohesion, chemical force, radiant force, and electricity . . .* (Wallace 1889, 476)

There seems to be little doubt that Wallace's intentions in the final section of his book mirror the earlier opinions of Boole, particularly those Boole expresses in the final chapter of his own book, *An Investigation of the Laws of Thought*:

> *If the mind, in its capacity of formal reasoning, obeys,*
> *whether consciously or unconsciously, mathematical*
> *laws, it claims through its other capacities of sentiment*
> *and action, through its perceptions of beauty and of*
> *moral fitness, through its deep springs of emotion and*
> *affection, to hold relation to a different order of things.*
> (Boole 1854, 444)

The nominal alliance in thought between Wallace and Boole
is in stark contrast to the intellectual connection established
between Darwin and Boltzmann on similar topics. Boltzmann,
an ardent Darwinian and unrepentant materialist, found
arguments like those of Boole and Wallace esoteric and
muddleheaded.

> *Thus it may happen to the mathematician that he,*
> *always occupied with his equations and dazed by their*
> *internal perfection, takes their mutual relationships for*
> *what truly exists, and that he turns away from the real*
> *world.* (Boltzmann [1890] 1973, 24)

For Boltzmann, mathematics was merely one useful representational system, and not a source of ultimate truth. Certain
areas of mathematics have been retained due to their success
at explaining the world not because they express an unfathom-
able platonic or spiritual reality. Boltzmann expressed what
we might now think of as an evolutionary epistemology—one
where ideas, in the struggle for explanatory preeminence,
are either retained or discarded. Boltzmann went so far as to
describe the nineteenth century as the century of Darwin and
ruefully thought of himself as the Darwin of physics.

Unlike Wallace, and like Darwin, Boltzmann had no qualms
about positing continuities in all traits between humans and
non-humans:

> *The analogy between the sensations of man with those*
> *of the highest animals is so perfect that we absolutely*
> *must ascribe objective existence to the latter sensations.*
> (Boltzmann 1897, 30)

And Boltzmann had little patience for the immutability and perfection of logic as described by Boole:

> *In the history of science there are many cases where the-*
> *orems were either proved or refuted through evidence*
> *which was thought to correspond to laws of thinking,*
> *while now we are convinced of its futility.* (Boltzmann
> [1899] 1973, 38)

Darwin's influential writings on the intellectual faculties and their selective value made the case for the evolutionary benefit of the intellect in his book *The Descent of Man:*

> *These faculties are variable; and we have every reason to*
> *believe that the variations tend to be inherited. Therefore,*
> *if they were formerly of high importance to primeval*
> *man and to his ape-like progenitors, they would have*
> *been perfected or advanced through natural selection. Of*
> *the high importance of the intellectual faculties there can*
> *be no doubt, for man mainly owes to them his preëmi-*
> *nent position in the world. We can see, that, in the rudest*
> *state of society, the individuals who were the most saga-*
> *cious, who invented and used the best weapons or traps,*
> *and who were best able to defend themselves, would rear*
> *the greatest number of offspring.* (Darwin 1871, 153)

Henri Poincaré, much like Ludwig Boltzmann, was an admirer of Charles Darwin and was evidently influenced by his style of naturalistic reasoning. In his philosophical essays, Poincaré made arguments that were rather similar to those of Boltzmann, explaining the appeal of mathematical theories in material terms: their origin in empirical observation, their ability to encode the regular structure of observation, and through correspondence with reality demonstrate significant utility.

> *Whence comes this concordance? Is it merely that things which seem to us beautiful are those which are best adapted to our intelligence, and that consequently they are at the same time the tools that intelligence knows best how to handle? Or is it due rather to evolution and natural selection?* (Poincaré [1903] 2017, 23)

COMPLEX TIME

French scholars tended to resist Darwin's works for either nationalistic reasons (favoring Jean-Baptiste Lamarck and George Cuvier) or on grounds of its irreligiousity (Farley 1974). Germany was far more receptive having been primed for transmutation by both Immanuel Kant and Johann Wolfgang von Goethe and championed by Ernst Haeckel (Richards 2013). Most prominent among Darwin's more formidable critics were two of the founders of statistical mechanics in Britain, James Clerk Maxwell and William Thomson (Lord Kelvin) (Burchfield 2009). Their criticisms are of particular interest because they represent an attempt to establish a direct connection between physical and biological theories: an effort to establish a mechanistic incompatibility between physical law and two of Darwin's assumptions: (1) the time required for natural selection to have evolved complex forms of life; and (2) the mechanism of inheritance required for selection to operate. Unlike Wallace's criticisms, which might be ascribed to a lack of imagination, Maxwell and Thomson's criticisms were full of scientific ingenuity, rigor, and represent a generative clash of paradigms that continued into the twentieth century as complexity continued its materialization.

There is no procedure in Darwin's published work for estimating the time required for a given increment in evolution to take place. Darwin was writing before dendrochronology had become a calibrated technique for chronometry (interestingly, Charles Babbage used dendrochronology to estimate the age of trees in peat bogs in *The Ninth Bridgewater Treatise* in 1938). There was no radio-carbon dating (1940s–) and no techniques of phylogenetic inference (1950s–). There was evidence from the fossil record and techniques of stratigraphic superposition. Darwin had discussed the imperfections of the fossil record in chapter 10 of *The Origin of Species* and concluded that it could not be relied upon for any kind of accurate estimates. Lyell in the *Principles of Geology* summarized the logic of stratigraphy in terms of the relationships between age and vertical position, unconformities, erosion, weathering, and uniformitarianism. By means of which Lyell had estimated the earth's age to the order of a hundred million years old.

Starting in the 1860s, Lord Kelvin made a series of calculations in which he estimated the age of the earth based on the radiative life-span of the sun, heating of the earth by tidal friction, and geochronological calculations. Kelvin's geochronology assumed the nebular hypothesis of planet formation: start with a molten ball of uniform temperature and subject it to cooling. In 1863 Kelvin's estimates of a hundred million years were in accordance with those of Lyell. By the 1890s, these estimates had been reduced to twenty million based in part on an estimate of the sun's age at twenty million.

Despite the lack of a theory of evolutionary time, Darwin considered these estimates damaging to his position, going so far as to write to one correspondent, "I am greatly troubled at the short duration of the world according to Sir W. Thompson, for I require for my theoretical views a very long period *before* the Cambrian formation." (F. Darwin and Seward 1903, 164)

Darwin recruited his son George (Fellow of Trinity College Cambridge) to his cause, asking for his help in reanalyzing Lord Kelvin's estimates. These calculations demonstrated that tidal friction would have considerably slowed the cooling of the earth. When used in combination with the observations of John Perry on the correct estimation of geothermal gradients at the earth surface based on interior convection (rather than Kelvin's diffusion), and buttressed by the discovery of a molten core, the estimate of the age of the earth was in the billions. Charles Darwin crowed in a letter to George on October 29, 1878, "Hurrah for the bowels of the earth & their viscosity & for the moon & for all the Heavenly bodies & for my son

George." (Mason 1994, 119) We now estimate the age of the earth at approximately four and half billion years.

James Clerk Maxwell had a rather different objection to Darwin's theories. Maxwell saw flaws in Darwin's theory of inheritance, the so-called theory of pangenesis. Darwin's idea was that every cell in the body shed smaller undifferentiated cells (gemmules) into the bloodstream. Gemmules migrated into the germ line, constituting a memory of a generation, thence to be transmitted through reproduction to offspring. Each offspring blends the gemmules of its parents, thereby supporting the adaptive correlations within a lineage. We now know this theory to be incorrect based on techniques of genetics and cell biology.

Maxwell's objections were based on reasoning derived from his kinetic theory of heat (Maxwell 1871, 42):

> Some of the exponents of this theory of heredity have attempted to elude the difficulty of placing a whole world of wonders within a body so small and so devoid of visible structure as a germ, by using the phrase struc-tureless germs. Now, one material system can differ from another only in the configuration and motion which it has at a given instant. To explain differences of function and development of a germ without assuming differ-ences of structure is therefore to admit that the properties of a germ are not those of a purely material system.

Maxwell's observations were not as quantitative as Lord Kelvin's but proved to be more lasting. It was not until the publication of Erwin Schrödinger's *What Is Life? The Physical Aspect of the Living Cell* in 1944 that the basis of "differences in structure" was ascribed to an aperiodic crystal. Like Maxwell before him, Schrödinger sought to reconcile the stability of inheritance with the known facts of statistical mechanics. It is fair to say that to this day the tension between biological form and function and physical theory remains an extremely generative connection.

The creators of statistical mechanics, evolutionary theory, formal logic, dynamics, and computation were members of a scholarly community connected by a dense web of institutions and societies. Through both publications and meetings they were influenced by shared cultural values and by each other's research results. Many of their ideas originated in the effort to understand the properties of organisms and machines. All of them struggled to explain the origin of stable, efficient, predictable, and practical technologies—both designed and evolved. It would take the twentieth century to commence their integration. And accompanying this synthesis, a reduced emphasis on physics and an increasing dependence on computational principles, which disclosed the information schemas supporting complex reality.

On the Origin of Design

One challenge for science at this time was that many machine-age concepts lived an anthropomorphic existence: mechanical function was used to make sense of the domains of purposeful traits and behaviors in society. And by the metaphysical norms of the period their origin was presumed to lie squarely within the purview of God's design extending subcreatively into human lives. This is a perspective much at odds with contemporary thinking (paradigmatically so) and is often forgotten in our ahistorical endeavors to construct an unbroken narrative linking sequential epochs in scientific belief.

The challenge was therefore to naturalize mechanics, generalize its insights into mechanical work and natural history, and do so respecting the postulates of natural theology. Much of this effort was accomplished through ideas circulating after the publication of a hugely ambitious scholarly project.

There are few nineteenth-century publications more ambitious in drawing parallels between the designed and the natural by pursuing evidence for the "teleological thesis" than the eight-plus-one *Bridgewater Treatises*. The first eight volumes were published between 1833 and 1836 and now stand as rather inspired tracts on mechanics suspended in an incongruous distillate of apologetics (Topham 2022).

The Ninth Bridgewater Treatise was an unauthorized contribution to the series published by Charles Babbage in 1837, proposing a new theory of "miracles" achieved by reconciling natural law, contingency, and computation. The first treatise was published in two volumes by Thomas Chalmers, establishing the logic and tone of the whole series: building a rhetorical connection between scientific principles—including parsimony—the new mechanical machines, and design.

Now it is a commonly received, and has indeed been raised into a sort of universal maxim, that the highest property of wisdom is to achieve the most desirable end, or the greatest amount of good, by the fewest possible means, or by the simplest machinery. When this test is applied to the laws of nature—then we esteem it, as enhancing the manifestation of intelligence, that one single law, as gravitation, should, as from a central and commanding eminence, subordinate to itself a whole host of most important phenomena; or that from one great and parent property, so vast a family of beneficial consequences should spring. And when the same test is applied to the dispositions, whether of nature or art— then it enhances the manifestation of wisdom, when some great end is brought about with a less complex or cumbersome instrumentality, as often takes place in the simplification of machines, when, by the device of some ingenious ligament or wheel, the apparatus is made equally, perhaps more effective, whilst less unwieldy ...
(Chalmers 1833, 49–50)

The most far-reaching and complete vision of design in nature, adduced through analogy with machines, was expressed in the third Bridgewater Treatise by William Whewell. Whewell describes mechanisms at many scales, from the global atmospheric system to the motion and stability of the solar system:

The contemplation of the atmosphere as a machine which answers all these purposes, is well suited to impress upon us the strongest conviction of the most refined, far-seeing, and far-ruling contrivance. It seems impossible to suppose that these various properties were so bestowed and so combined, any otherwise than by a beneficent and intelligent Being, able and willing to diffuse organization, life, health, and enjoyment through all parts of the visible world . . . (Whewell 1833, 127–128)

We know at present very little indeed of the construction of this machine. Its existence *is, perhaps, satisfactorily made out; in order that we may not interrupt the progress of our argument, we shall refer to other works for the reasonings which appear to lead to this conclusion. But whether heat, electricity, galvanism, magnetism, be fluids; or effects or modifications of fluids; and whether such fluids or* ethers *be the same with the luminiferous ether, or with each other; are questions of which all or most appear to be at present undecided, and it would be presumptuous and premature here to take one side or the other* (Whewell 1833, 139–140).

The Bridgewater Treatises would largely be of historical interest, and not a core concern for the origins of complexity, if they had not proven to be of such importance to its prefoundational architects, including Charles Darwin, Alfred Russel Wallace, James Clerk Maxwell, Robert Boole, and Charles Babbage. *The Bridgewater Treatises* were demonstrably of direct influence on science, but one should not underestimate the order they lent to the zeitgeist of natural science.

The larger narrative of *The Origin of Species*, Darwin's so-called "one long argument," was a naturalistic account of the origin of novelty or species, and was strongly influenced by Darwin's repeat readings of Whewell and his natural theology. This influence extended to a quote from Whewell as the opening epigraph of *The Origin of Species*:

"But with regard to the material world, we can at least go so far as this—we can perceive that events are brought about not by insulated interpositions of divine power, exerted in each particular case, but by the establishment

of general laws" William Whewell, Astronomy and General Physics Considered with Reference to Natural Theology (Darwin 1859, epigraph)

Darwin used Whewell's system as a pretext to invoke Newton's gravitational law as analogy and widely acclaimed, authoritative prequel, for the generality of natural selection presented in the third edition of *The Origin of Species.*

> *It has been said that I speak of natural selection as an active power or Deity; but who objects to an author speaking of the attraction of gravity as ruling the movements of the planets? Everyone knows what is meant and is implied by such metaphorical expressions; and they are almost necessary for brevity. So again it is difficult to avoid personifying the word Nature; but I mean by nature, only the aggregate action and product of many natural laws, and by laws the sequence of events as ascertained by us.* (Darwin 1861, 85)

Darwin chose the gravity metaphor as a mechanical metaphor for selection in 1859, but had agreed to Wallace's far more fitting suggestion of Watt's centrifugal governor in the first selection paper with Alfred Russel Wallace in 1858:

> *The action of this principle is exactly like that of the centrifugal governor of the steam engine, which checks and corrects any irregularities almost before they become evident; and in like manner no unbalanced deficiency in the animal kingdom can ever reach any conspicuous magnitude, because it would make itself felt at the very first step, by rendering existence difficult and extinction almost sure soon to follow.* (Darwin and Wallace 1858, 62)

Darwin had switched out an engineered mechanism of contingent generality for a law of universal application. By the dictates of natural theology, neither metaphor questioned the ultimate origin of design. But the choice of gravity pushed causality back into a primordial and parsimonious beginning, and suggests a far more expansive vision for selection than a practical regulator of engine speed.

It was a decade after the publication of Darwin and Wallace's paper, in 1868, that James Clerk Maxwell explained the logical basis of Darwin and Wallace's intuition for the balancing of the natural world, demonstrating under what conditions regulation or moderation would stabilize or destabilize a dynamical system. In so doing Maxwell invented the field of control theory, and established one of the foundational ingredients of complexity science—the principle of feedback control—the kernel of the selection process.

Maxwell himself was an avid reader of Whewell and, like Darwin before him, sought a means of reconciling physics with machines and theology. In one of the more extraordinary convolutions of the entwining of incompatible ideas, Maxwell speculated on the uniformity of the fundamental particles of nature in relation to the production-line in an article from 1873:

> *We are thus assured that molecules of the same nature as those of our hydrogen exist in those distant regions, or at least did exist when the light by which we see them was emitted. . . .* (Maxwell 1873, 375)

> *Each molecule, therefore, throughout the universe, bears impressed on it the stamp of a metric system as distinctly as does the metre of the Archives at Paris, or the double royal cubit of the Temple of Karnac . . .*

[...]

None of the processes of Nature, since the time when Nature began, have produced the slightest difference in the properties of any molecule. We are therefore unable to ascribe either the existence of the molecules or the identity of their properties to the operation of any of the causes which we call natural ... On the other hand, the exact equality of each molecule to all the others of the same kind gives it, as Sir John Herschel has well said, the essential character of a manufactured article, and precludes the idea of its being eternal and self-existent. (Maxwell 1873, 376)

Maxwell's observation, by the standards of our own science, strikes us as utterly surreal.

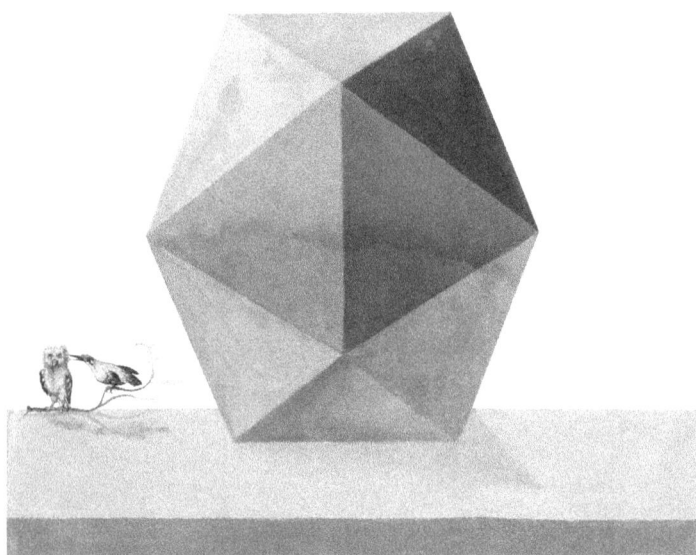

Definitional Pursuits

Outline of an Ontology

This section presents a chronology and distillation of papers and books that have elucidated and provided definitions for both the purview and principles of complexity science. These are all drawn from a *superset* of the *foundational papers*. They reflect in profound ways the four pillars of complexity established in the nineteenth century. In almost every case we can chart the two or more fields that they are seeking to integrate.

Definitional projects are not synonymous with foundational contributions; hence many important foundational papers do not appear in this section. Illustrative exclusions include those by Claude Shannon on communication and Alan Turing on computation. Both are incontestably foundational for complexity, yet neither sought to articulate the character of the complex domain. Both theoretical computer science and information theory provide important ideas and methods in complexity science while not being synonymous with it. After all, ideas from information theory and computer science are just as important to engineering classical machines and to fundamental physics as they are to complexity science. The ideas of Shannon and Turing, when supplemented by considerations from nonlinear dynamics, thermodynamics, or evolution, have informed more explicit definitions of complexity provided by Warren Weaver, Herbert Simon, John Holland, Stephen Wolfram, and others.

It is also often the case that new ideas from seemingly unrelated fields—those without connections to questions of complexity—prove to be important. The Brouwer fixed-point theorem, conceived as a means of characterizing the topology of Euclidean spaces, was the central analytical technique deployed by John Nash in developing his competitive game solution concept— the Nash equilibrium. The method is agnostic toward complexity and is a valuable pragmatic approach from applied mathematics.

Many recent efforts make use of power laws (Brown and West 2000), perturbation theory (Simmonds and Mann, Jr. 1986), and networks (Newman 2010). All introduce powerful frameworks and methods from physics and the social sciences to study complex systems. But these need not themselves elucidate principles of complexity. Typically they are important ingredients in more inclusive attempts at coming to terms with robust and sometimes near-universal patterns of organization. These include the use of scaling theory to analyze evolved metabolic networks and the use of network statistics to make sense of the structure of neural circuits or social structures.

Complex phenomena are not to be confused with methods; they are what the methods discover: we might describe these as the *syntax* (rules revealed by analysis) and *semantics* (functions revealed through analysis) ascribed to the results of the analysis. Thus a perturbation theory can be applied to derive a scaling relationship between mass and spin in particle physics or area and species diversity in ecology; it is only the latter by virtue of the relationships arising through the constrained optimization achieved via evolutionary history (near-countless frozen accidents fixed by selection) that benefits from the label complex.

The primary purpose of this section is to demonstrate two things: (1) the emphasis placed on revealing a new domain of inquiry, or an ontology, that does not fit naturally into any one existing disciplinary focus and thereby requires combining new ideas (many from the prefoundational period); and (2) the historical accumulation, or ratcheting-up, of principles required to adequately address this domain through an historical timeline. The more recent definitions of complexity build on principles provided in prior publications—not always explicitly. Complexity is an evolving field accumulating concepts; it is not a stationary pursuit, no more than biology, archaeology, or any other vibrant area of inquiry.

A Chronology of Definitions

The ongoing challenge for all of these definitional projects has been to arrive at concepts that are inclusive of physical processes but sufficiently extended to contend with extra-physical or adaptive phenomena (e.g., life and intelligence). It is very natural that some researchers place greater emphasis on the recruitment of physical laws with key modification (typically those coming from nonequilibrium physics), whereas others introduce entirely new concepts that they seek to connect—inter-theoretically—with established physical processes (those coming from biology, computer science, and the social sciences). Nonetheless, the overall impression is one of surprising coherence. In most definitions there is a primary thesis of stable nonsymmetric hierarchies (in both space and time) supporting transmissible, adaptive functions.

What we now think of as complexity science appears to have become organized into a loosely bound set of related principles (systems) governing the adaptive world between the 1940s and early 1990s. The earliest publications sought to differentiate fledgling complexity science (cybernetics, general systems theory, living systems theory, etc.) from physics. These consisted largely in definitions emphasizing out-of-equilibrium behavior, dissipative or irreversible dynamics, and self-regulating and self-reproducing systems. Later definitions place greater emphasis on integrating nonequilibrium dynamics into functional capabilities, including adaptation, exploration, robustness-related mechanisms, and computation.

These publications also emphasize the broad compass of complexity principles—that they could not be defined in any restricted relation to existing disciplinary boundaries. In this respect, the emerging paradigm not only introduced new principles, but is fundamentally iconoclastic with respect to existing scholarly institutions structured by disciplines. Researchers such as Norbert Wiener and Warren Weaver made this a very explicit part of their articulation of a new science.

Complexity science increasingly pursues scholarship structured by shared principles, forms, and functions. Through this new approach, domains that diverge when conditioned on discipline (e.g., economics and ecology or evolution and archaeology), came to be seen as variations on a common theme, or case studies, for similar patterns manifesting in different structures and at different space and time scales. A common feature of these contributions is that they simply do not see the disciplinary boundary as particularly meaningful. This is in contrast to more conventional interdisciplinary work that seeks to make progress by combining ideas from previous areas to solve a new problem, but maintain disciplinary distinctions as a justification for their novelty.

Extracting Principles

From the definitional table (tab. 2) we can extract from each study, in chronological order, a central principle and consider the implications of their combinations and conjunctions. After all, classificatory definition is never satisfying, and even a real definition that alludes to contingent material properties will be constrained by a field of inquiry and can often feel restrictive. A more useful approach is to provide an explication of shared areas of interest in order to arrive at a more exact understanding of the richness of the complex domain.

1940s

ROSENBLUETH, WIENER & BIGELOW:
BEHAVIOR, PURPOSE, AND TELEOLOGY

YEAR	NEW CONCEPTS	INHERITED CONCEPTS
1943	T	

A founding paper in cybernetics in which behaviorism (reinforced input–output system) and purpose (T = *teleology*) are realized through feedback in a dynamical system.

Cybernetics integrates earlier work from engineering, dynamical systems, and psychology. The paper contains the seeds for many ideas in contemporary cognitive science, including policies that minimize prediction errors, and extrapolation through active inference. Complexity is treated as a self-regulating dynamical system.

HAYEK: THE USE OF KNOWLEDGE IN SOCIETY

YEAR	NEW CONCEPTS	INHERITED CONCEPTS
1945	M	T

Hayek builds on economic insights from Adam Smith on the role of the "invisible hand" to construct systems of tacit knowledge for allocating scarce resources through *markets* (**M**). Hayek was one of the earliest proponents of complexity as collective knowledge generation through decentralized processes mediated by market-based coordination mechanisms. Hayek's ideas share with cybernetics an interest in regulation towards an efficient goal but its logic comes from evolution, psychology, and political economy.

WEAVER: SCIENCE AND COMPLEXITY

YEAR	NEW CONCEPTS	INHERITED CONCEPTS
1948	OC	T, M

Warren Weaver was the first to use the term complexity (or *organized complexity,* **OC**) as a means of identifying a domain of inquiry spanning biology, markets, societies, and polities. A colleague of Norbert Wiener's when both worked on electrical anti-aircraft systems—a professional connection that informed Weaver's understanding of purposeful, or teleological, dynamics. Weaver's ideas on **OC** are defined in distinction to both simplicity, understood as classical dynamics, and *disorganized complexity,* derived from a close reading of the work of Willard Gibbs and the statistical-mechanics literature. Weaver's understanding of science is influenced by Hayek's market-based reasoning concerning the division of labor required to make progress on complex problems.

WIENER: CYBERNETICS AND CONTROL

YEAR	NEW CONCEPTS	INHERITED CONCEPTS
1948	IT	T, M, OC

A monumental effort to establish cybernetics at "the boundary regions of science" by integrating ideas from information theory, control theory, physiology, cognition, computer science, and social science. Cybernetics seeks to analyze a broad range of human functions based on analogies to "mechanico-electrical systems." The central organizing principle is an expansive conception of *information transmission* (**IT**) in machines, brains, and societies. For Wiener, complexity and cybernetics are largely synonymous.

1950s

VON BERTALANFFY: OPEN SYSTEMS IN PHYSICS AND BIOLOGY

YEAR	NEW CONCEPTS	INHERITED CONCEPTS
1950	OS	OC

An approach to complexity stressing the importance of the physics of *open systems* (**OS**) as these manifest in all life-like phenomena. A synoptic investigation of time-independent steady states that do not correspond to a maximum of entropy and demonstrate an independence from initial conditions (equifinality). A project to connect classical dynamics and thermodynamics to augmented models suitable for life. Best seen as an effort to establish the necessary but not sufficient conditions for complexity of a kind considered by Wiener, Hayek, and Weaver.

MILLER: GENERAL THEORY OF BEHAVIORAL SCIENCE

YEAR	NEW CONCEPTS	INHERITED CONCEPTS
1948	LST	T, IT, OS

A descriptive framework in which life is presented as an hierarchically structured organization of increasingly open systems. The lowest levels are described by equilibrium physical principles and subsequent inclusive levels by far-from-equilibrium physics and organic reactions. The highest levels are characterized by functional mutual dependencies among evolved structures characteristic of life. *Living systems theory* (**LST**) asserts

Table 2. The core definitional ingredients of complexity science slowly begin to coalesce in the mid-twentieth-century. Through a series of highly influential books and papers, the complex domain comes into focus. Through time, an implicational scale (a series of entailments) is built up, culminating in an expansive definition for a phenomenal world of self-organizing and selected adaptive patterns.

REFERENCE/DEFINITION	ONTOLOGY/ DOMAIN	STATED FIELD
ROSENBLUETH, WIENER, AND BIGELOW (1943): Dynamical systems with active or purposeful behavior; directed towards the attainment of a goal; "voluntary activity" supported by feedback and prediction (extrapolative)	Biology, engineering	"Philosophy of science"
HAYEK (1945): Systems planning through local/decentralized knowledge of "time and place": markets and the price system; spontaneous sources of order (catallaxy; 1978)	Resource allocation in the economy and society	"Economics"
WEAVER (1948): Systems of organized complexity; sizeable numbers of factors which are interrelated into an organic (functional) whole	Biology; society; economy	"Complexity science"
WIENER (1948): Systems controlling functional information through feedback to achieve homeostasis; stability and goal-seeking	Biology; society; engineering; culture	"Cybernetics"
VON BERTALANFFY (1950): Open systems; dynamic equilibrium; efficient metabolism; principle of equifinality; thermodynamics of irreversibility (building on Prigogine 1947)	Biology, chemical engineering	"General systems theory"
MILLER (1955): Tripartite concrete, conceptual, and abstract systems; structured subsystems; living processes; relationships between growth, decay, and termination	Biology; organizations; nations	"Living systems theory"

REFERENCE/DEFINITION	ONTOLOGY/ DOMAIN	STATED FIELD
SIMON (1962): Systems whose whole is more than the sum of its parts; hierarchy; evolution; functional decomposition; self-reproducible	Society; biology; physics; chemistry	"Behavioral science," "complexity science"
HOLLAND (1962): Adaptive systems; adaptation as the generation of programs that address environmental distributions of problems; feedback operating through differential activation	Biology; computation; logic	"Complexity science"
CONANT AND ASHBY (1970): Regulatory systems; feedback control; prediction, inference; and complexity of controller matching that of environment	Engineering; neuroscience	"Systems science," "control theory"
ANDERSON (1972): "More is Different": complex systems dominated by broken symmetries; reductionism does not imply constructionism: each level requires a different conceptual structure	Many-body theory; chemistry; biology; culture	"Complexity science"
VON FOERSTER (1972): A theory of "observers": "description- invariant subjective worlds"; representational systems; recursive and self-referring systems	Living systems	"Epistemology of living things"
SIMON (1973): Near-decomposable systems; evolvable hierarchies; alphabet-based; systems of languages and programs	Biochemistry; biology; engineering; society	"Complex systems"
VARELA, MATURANA, AND URIBE (1974): Unitary and autopoietic organizations; recursive networks of unitary processes supporting mechanisms of self-production; non-determinant elements; autonomous systems	Living systems	"Autopoiesis," "complex systems"

REFERENCE/DEFINITION	ONTOLOGY/ DOMAIN	STATED FIELD
HAKEN (1977): Self-organized structures emergent from chaos; structures induced by fluxes of energy and matter; biological function combining dissipative and non-dissipative structures; the "slaving principle"	Statistical physics; chemistry; biology; society	"Synergetics"
PRIGOGINE AND ALLEN (1982): Dissipative dynamical systems; symmetry breaking; far from equilibrium; long-range correlations; selection and algorithmic complexity	Dynamical systems; statistical physics; pattern formation	"Complexity"
WOLFRAM (1984): Discrete spatially extended dynamical systems with non-linear cooperative effects (rule systems); spatial computation; complexity and non-computability of rule systems	Dynamical systems; emergence; pattern formation; computation	"Complex systems"
KAUFFMAN (1990): Systems at the edge of chaos; multi-peaked fitness landscapes; replication as autocatalysis; evolvability as selective meta-dynamics	Dynamical systems; evolution; gene regulation; development	"Complex systems"
FARMER (1990): Connectionism; networks of interacting agents with time-varying— adaptive—connection strengths	Neural networks; classifier systems; immune networks; autocatalytic networks	"Complex systems"
GELL-MANN AND HARTLE (1994): Information gathering and utilizing systems (IGUS); the intersection of computational devices with physical reality to create quasiclassical realms	Society; biology; physics; chemistry	"Complex systems"

that organisms are built outwards from a simple core to a complex periphery. The core is populated by chemical elements, or basic organisms, and the periphery by *larger organisms*, including cells. An effort to defy the *continuum hypothesis* of the disciplines that tend to dichotomize ontology and advocate for a complex spectrum of life.

1960s

SIMON: THE ARCHITECTURE OF COMPLEXITY

YEAR	NEW CONCEPTS	INHERITED CONCEPTS
1962	ND	T, M, OC, IT, OS

Simon proposes after Weaver a second explicit definition for complexity science. Simon extends organized complexity into a focus on the value of nested spacetime structures. Complex structures are characterized as *near-decomposable* (**ND**) hierarchies. These resemble Miller's description of living systems, with a focus on robust assembly, the importance of a separation of time scales, information-processing layers, and greater comprehensibility through modularity.

HOLLAND: ADAPTIVE SYSTEMS

YEAR	NEW CONCEPTS	INHERITED CONCEPTS
1962	AS	T, OC, IT

Generalizing the Darwin–Fisher conception of an *adaptive system* (**AS**) into a broad range of logical processes described by automata theory. Formally, each adaptive agent is a generative procedure running on a universal computer to solve problems posed by the environment. Evolution is conceptualized as the interface of a universal generation procedure outputting populations of programs that solve contingent problems. A complex system defined as an abstract form of life characterized by information describing an adaptive program.

1970s

CONANT AND ASHBY: GOOD REGULATORS

YEAR	NEW CONCEPTS	INHERITED CONCEPTS
1970	GR	T, OC, IT, AS

Complex systems as models of their environment, or *good regulators* (**GR**). Building on ideas from control theory, information theory, and psychology in order to open up the black box (behaviorism) of adaptive feedback. Hypothesizing that all complex systems need to have "agentic" features—an internal model—when dealing with nontrivial environmental dynamics. The brain is described as the prototypical complex system that models the world around it in order to promote control and thereby survival.

ANDERSON: MORE IS DIFFERENT

YEAR	NEW CONCEPTS	INHERITED CONCEPTS
1972	E	OC, IT, OS, AS

Introduces the idea that, for systems at any reasonable meso-scopic scale, symmetry-breaking abrogates the power of the fundamental laws of physics. This implies that effective theories are not merely useful approximations of reality but strictly necessary to model collective phenomena— *emergence* (**E**). Emergence is therefore a signature of complexity and a profound justification for a pluralism of theories and of understanding. An inescapable refutation of ontological reductionism based on many-body physics. Complex systems as increasingly complicated histories of broken symmetries.

VON FOERSTER: EPISTEMOLOGY OF LIFE

YEAR	NEW CONCEPTS	INHERITED CONCEPTS
1972	SB	OC, IT, OS, AS

Introduces a conceptual framework for exploring a recursive extension of Conant and Ashby through the idea of *self-observation* (**SB**). How to pursue an objective or description-invariant account for the subjective world; or an epistemology true to the living world. Building on ideas from logic and computation rooted in the ideas of Gödel and Turing, psychology derived from Piaget, and the linguistic philosophy of Wittgenstein. Complex systems as self-aware, reflexive, or, in the language of Douglas Hofstadter, "strange-loops."

SIMON: ORGANIZATION OF COMPLEXITY

YEAR	NEW CONCEPTS	INHERITED CONCEPTS
1973	SF	OC, IT, OS, ND, AS, E

Building on his own ideas of near-decomposability and hierarchy from 1962, and extending these to emergent hierarchies of programs that are *sealed-off* (**SF**) from lower levels. This is an emergent approach to computation that uses the idea of SF to justify a hierarchy of *effective theories* within a single system. Introduces an informal theory of types through level-dependent alphabets and languages. Complex systems as nested information-processing systems or hierarchies of formal languages.

VARELA, MATURANA, AND URIBE:
AUTOPOIETIC ORGANIZATION

YEAR	NEW CONCEPTS	INHERITED CONCEPTS
1974	AU	OC, IT, OS, ND, AS, E

Combining ideas of emergence, self-modeling, adaptation, universal construction, and computation to formalize an idea of *autopoiesis* (**AU**). An autopoietic system is a dynamical network of interacting components capable of self-synthesis, self-reproduction, and autonomy. Autonomy is a degree of independence from the environment, resembling contemporary ideas of the *Markov blanket* in Bayesian networks. A complex system is an autonomous open system that transcends its materials to synthesize a robust replicate.

HAKEN: SYNERGETIC SELF-ORGANIZATION

YEAR	NEW CONCEPTS	INHERITED CONCEPTS
1977	SO	OC, OS, AS, E

The application of ideas from the theory of phase transitions and bifurcation theory to explore a potentially universal class of synergetic *self-organizing* (**SO**) systems. In practice, focusing on a variety of nonequilibrium collective dynamics under the control of a relatively small number of internal or external control parameters. Self-organization is an important ingredient of complex systems. It is difficult to distinguish between the spontaneous emergence of ordered states and those resulting from prolonged evolutionary histories.

1980s

PRIGOGINE AND ALLEN: DISSIPATIVE SYSTEMS

YEAR	NEW CONCEPTS	INHERITED CONCEPTS
1982	DS	OC, OS, E, SO

The importance of the arrow of time in contributing to far-from-equilibrium organization or *dissipative structures* (**DS**). A cognate to synergetics, with a strong focus on simple physico-chemical processes that span both biotic and abiotic phenomena. An emphasis on the dynamics of irreversible processes giving rise to an arrow of time. Complexity emerges through self-organization understood in terms of long-range order and symmetry-breaking sensitively coupled to boundary conditions. Complexity theory as a theory of patterning, not a theory of function.

WOLFRAM: RULE SYSTEMS

YEAR	NEW CONCEPTS	INHERITED CONCEPTS
1984	RS	T, OC, IT, E, SO

A science of self-organizing pattern formation based on discrete, spatially extended dynamical systems. An analysis of both the rules governing these dynamics and different classes of spacetime patterns with varying degrees of computational power. Complexity as an emergent form of spatial computation achieved through collective dynamics of simple *rule systems* (**RS**).

1990s

FARMER: ADAPTIVE CONNECTIONISM

YEAR	NEW CONCEPTS	INHERITED CONCEPTS
1990	AC	TT, OC, IT, AS, E, SO

Complex systems as networks of adaptive agents that achieve a separation of time scales between slowly evolving connectivity and quickly changing coupling strength. The identification of common structures and functions in autocatalytic chemistry, nervous systems, immune systems, and ecosystems suggestive of a universal class of *agentic connectionist* (**AC**) system.

KAUFFMAN: EDGE OF CHAOS

YEAR	NEW CONCEPTS	INHERITED CONCEPTS
1993	EC	T, OC, IT, OS, AS, E, SO

A project to integrate selection and self-organization by building on structural correspondences between physical energy landscapes and biological fitness landscapes. A common feature of ordered states in nonequilibrium physical systems and functional states in adaptive systems is a tuning to the *edge of chaos* (**EC**). Complex systems as a broad class of phenomena that exist at the edge of chaos in order to more effectively catalyze reactions, form stable patterns, store memories, and ensure continued adaptability.

GELL-MANN AND HARTLE:
INFORMATION GATHERING AND UTILIZING SYSTEMS

YEAR	NEW CONCEPTS	INHERITED CONCEPTS
1994	IGUS	T, OC, AS, GR

Complex systems as *information gathering and utilizing systems* (**IGUS**). Any evolved or engineered system capable of adaptive and predictive information processing. The IGUS establishes a principled basis for the definition of what were once thought to be physics concepts: past, present, and future. Through the lens of the IGUS these three temporalities require complexity-based definitions. These are in the form of agent-centric computations embedded in Minkowski space whereby each agent experiences a unique world line.

Explicating Complexity

This sparse chronology serves to demonstrate both the specificity and the generality of definitions proposed for complex phenomena. Each definition might conveniently be placed along a spectrum in order to explicate complexity in relation to a continuum. Toward one end are those approaches exploring the reach of an extended physical theory of *open systems* (**OS**) dominated by *dissipative structures* (**DS**) and *self-organization* (**SO**), all displaying simple forms of *sealed-off* (**SF**) *emergence* (**E**). And toward the other end, systems whose *teleonomic* (**T**) features need to be accounted for in terms of *adaptive systems* (**AS**) producing *organized complexity* (**OC**) in a variety of *autopoietic* (**AU**) organizations capable of modeling the environment through *good regulators* (**GR**), which in some cases achieve *self-observation* (**SB**).

These definitions suggest that the distinction sometimes made between *complex systems* (**CS**) (non-evolved) and *complex adaptive systems* (**CAS**) (evolved), is too dichotomized. Very few systems of interest outside of idealized mathematical models or computer simulations conform to either of these two strict limits. It is more straightforward to describe all these systems as *complex systems*, recognizing that any constructed systems—evolved or engineered—will need to exploit far-from-equilibrium processes—and will fall somewhere in the interval described by a history of selected broken symmetries. *Complexity science is from this perspective a family of models and theories, all of which seek to explain the emergence of far from equilibrium structures, which vary in their adaptive capacity and length of evolutionary histories.* Taking our lead from both James Clerk Maxwell and Carl Linnaeus, we might describe the full complexity science spectrum as the genus (the domain of inquiry) and each point on the continuum as the species (particular processes required by the explanandum).

ENGINEERING EMERGENCE

There have been efforts to discover non-living phenomena at the CS limit: to find systems that self-organize into nontrivial ordered states spontaneously without an evolutionary history. Consider one of the favorite reactions adduced as a candidate: the *Belousov–Zhabotinsky* (**BZ**) reaction (Miyazaki 2013). The BZ reaction comprises around eighteen distinct chemical reactions organized into two autocatalytic cycles that are able to produce nonlinear chemical oscillations. This is an open system, showing a simple form of organized complexity, based on dissipative reactions, manifesting a simple form of hierarchy in the form of self-similarity—fractality. At present, the BZ reaction has only been observed in the laboratory when engineered by purposeful chemists and maintained under very exacting conditions. Candidate BZ reactions have been hypothesized to explain the growth patterns of evolved life-forms including for amoebae and slime molds. The BZ is in other words a complex system made by complex adaptive systems. And the same argument applies to Turing patterns, and to a rather broad range of non-equilibrium structures not yet observed unengineered in the abiotic universe.

A far more compelling candidate for a limit-case complex system is a spin-glass (Stein and Newman 2013).[5] A spin-glass is a disordered magnetic material. The magnetic moments of a spin glass freeze/quench into orientations that are arbitrary or random as a result of a mixture of positive and negative interactions among atoms. Unlike simple materials, spin-glasses show significant levels of frustration (unsolvable constraints) generating complicated energy landscapes with many metastable states. When

---

[5] Stein and Newman's monograph *Spin Glasses and Complexity* (2013) builds up from the physics of simple magnets to spin glasses and their applications. It is notable for its engagement with the history of ideas of complexity, including the limitations of physical models and frameworks applied to the complex domain.

moved out of equilibrium (e.g., when placed in magnetic fields for some duration) they have been shown to possess aging and memory properties. These are all features of engineered and evolved systems. This is why spin-glass models (typically built on the Sherrington–Kirkpatrick mean field limit of the Edwards–Anderson model) provide a very powerful framework for studying a range of complex systems.

Daniel Stein and Charles Newman have addressed the complexity status of spin-glasses directly. As regards the use of *spin-glass models* (**SGM**) as *mathematical tools*: they are useful for analyzing phenomena as diverse as computational complexity, protein-folding, content-addressable memory, and social dynamics. SGM, much like empirical networks abstracted into matrices or biological growth abstracted into systems of differential equations, have proven to be an important tool in the analysis of complex systems. As regards *spin-glasses* (**SG**) as *physical systems* (e.g,. dilute magnetic alloys), Stein and Newman propose that:

> *spin glasses deserve at least the rubric of quasi-complex system. Whether truly complex or not, they have provided mathematical descriptions of important aspects of complexity . . .* (Stein and Newman 2013, 237–238)

Any rich field has at its boundary an escarpment not a wall. The SG forces us to reckon with mechanisms that we might wish to position closer or further from the center of any inquiry into complexity.

How to Think about Post-Evolutionary Physics

Applications of both equilibrium and non-equilibrium physics principles to complex systems often consist in the applications of variational principles to existing adaptive structures. The physics of adaptive matter is the search for parsimonious principles that might coexist with, or be placed above, unparsimonious mechanisms. A few examples worth analyzing in a little more detail include theories of development and morphology and theories of metabolic scaling. All of these pursue a minimal approach to effective theories, whose screened or sealed off constituent parts, are contingent forms of adaptive matter. In this respect, these are all kinds of post-evolutionary physics: theories of emergent properties that place evolution in a fundamental position relative to the variational principle—an inversion of the usual sense of fundamental.

D'ARCY WENTWORTH THOMPSON
ON GROWTH AND FORM

There are foundational works inimical to adaptive complexity in a philosophical sense: ideas like those in D'Arcy Wentworth Thompson's 1917 book *On Growth and Form*. This is very much a self-styled work of biophysics, seeking to reduce complexity by focusing on structural phenomena dominated by physical laws while deprecating questions of adaptation and function. The importance of *On Growth and Form* was to demonstrate that a significant degree of functional variation ascribed to selection could be derived from variational principles. It is in this way crucial to our understanding of complexity despite its ambition to eliminate it.

> *How far, even then, mathematics will* suffice *to describe, and physics to explain, the fabric of the body no man can foresee. It may be that all the laws of energy, and all the properties of matter, and all the chemistry of all the colloids*

*are as powerless to explain the body as they are impotent
to comprehend the soul. For my part, I think it is not so.*
(Thompson 1917, 8)

There are several notable chapters where Thompson challenges
the selective account of a biological structure. In "A Note Upon
Torsion," Thompson explains the twining of a stem about its
own axis as a:

*temporary adhesion or "clinging" between it and the
growing stem which twines around it; and a system of
forces is thus set up, producing a "couple," just as it was
in the case of the ram's or antelope's horn through direct
adhesion of the bony core to the surrounding sheath. The
twist is the direct result of this couple, and it disappears
when the support is so smooth that no such force comes to
be exerted.* (Thompson 1917, 626)

Darwin explains the same phenomenon thus:

*The stem probably gains rigidity by being twisted (on the
same principle that a much twisted rope is stiffer than a
slackly twisted one), and is thus indirectly benefited so as
to be able to pass over inequalities in its spiral ascent, and
to carry its own weight when allowed to revolve freely.*
(Darwin 1891, 10)

There is no conflict between these accounts. We resolve each
of them into proximate and ultimate explanations. Thompson
constructed a nice physical rule that in Darwin's estimation
achieves a selective reward. Despite the compatibility of these
accounts, debates in ink like the one between Thompson and
Darwin continue to this day, typically because the variational
argument fails to recognize that it has been made possible by a
very long evolutionary history. That is to say, elegant forms of
self-organizing are given license through rather messy genetic
processes.

NICOLAS RASHEVSKY ON TOPOLOGY

Following in Thompson's footsteps, Nicolas Rashevsky proposed a biophysics based on mathematical and fundamental physical principles including the principles of diffusion, drag forces described using hydrodynamics, and topological theorems.

The abstract of Rashevsky's paper on "The Geometrization of Biology" (1956) is worth reproducing in full:

> *The twentieth century has witnessed a geometrization of physics, that is, a reduction of the basic concepts of physics to geometric concepts. The topological approach to biology, recently proposed and to some extent developed by the author, is a small step in the direction of geometrization of biology, but is unable to achieve the main purpose of such a geometrization of biology, namely, the reduction to geometric concepts of such purely biological concepts as ingestion, digestion, assimilation, etc. To achieve this purpose we must find geometric structures or spaces, in which different geometric properties stand to each other in the same formal logical relation, as the different concepts of biology stand to each other. If this were possible, then a set of geometric theorems could be "translated" by an appropriate "glossary" into a set of biological laws.* (Rashevsky 1956, 31)

It goes without saying that Rashevsky's contributions to current understanding of ingestion and digestion are negligible. This is not because topological considerations are without merit. The problem is the topological principles were not adequately constrained by the evolved physiology of life forms. If Rashevsky had understood his own post-evolutionary physics, constraining models by common anatomy and physiology, one suspects his ideas would have proved far more influential.

GEOFFREY WEST AND JAMES BROWN
ON ALLOMETRY

At the opposite end of the spectrum is metabolic scaling theory. This is a theory that makes selection fundamental and then pursues surprisingly universal consequences. The challenge is to explain the widespread observation that macroscopic features of organisms and ecosystems scale as powers of one-quarter.

Geoffrey West and James Brown (2005) explain this regularity as the outcome of natural selection pushing up against the bounds of energetic constraints:

> We have proposed a set of principles based on the observation that almost all life is sustained by hierarchical branching networks, which we assume have invariant terminal units, are space-filling and are optimized by the process of natural selection. We show how these general constraints explain quarter power scaling and lead to a quantitative, predictive theory that captures many of the essential features of diverse biological systems. (West and Brown 2005, 1575)

The metabolic scaling theory is a sophisticated demonstration of the full complexity spectrum. The theory rests on a teleology of adaptation, a functional hierarchy of circulation, dissipative dynamics of resource allocation, and self-organization during the formation of networks. And the theory relies on emergent properties of fractal networks, which permits the derivation of a biological effective theory, captured in terms of the physical dimensions of space. Through evolution, one of the most fundamental features of the universe—spatial dimensionality—comes to play a central role in adaptation. One might say that scaling theory shows us how evolution has discovered physics.

Synoptic Surveys

Complexity science has been well served by books and monographs that both review and analyze the field. Despite the different scholarly provenances of authors, there is substantial overlap in interests and topics in many books.

A clue to the unity of complexity science is how many of the books explaining complexity science have very similar chapter headings. Table 3 is a representative list organized chronologically.

Nearly all of the books discuss chaos, cellular automata, social insects, nervous systems, immune systems, brains, and markets. And all of the authors make a form of emergence the central organizing concept. This agreement is achieved despite differential emphasis on the role of noise, adaptation, and determinism in the emergence of physical phases, dynamical patterns, or biological functions.

There is a slight chronological trend in which principles and ontology are emphasized in the earlier books and methods and models in the later books. Presumably this reflects some combination of the maturation of the field and a growing pressure from society to demonstrate the practical utility of complexity science.

Table 3. A small sample of books that set out to provide a synoptic introduction to principles bearing on complexity science or its adjacent domains. These are organized by publication date spanning the last four decades.

AUTHOR(S)	TITLE/YEAR	ABBREVIATED DEFINITION
DOUGLAS R. HOFSTADTER	*Gödel, Escher, Bach: An Eternal Golden Braid (1979)*	Living systems as pattern-forming mechanisms for encoding the world and themselves through recursion ("strange loops")
MANFRED EIGEN AND RUTHILD WINKLER	*Laws of the Game: How the Principles of Nature Govern Chance (1982)*	Systems capable of exploiting intrinsic sources of randomness to produce emergent forms of combinatorial organization
ILYA PRIGOGINE AND ISABELLE STENGERS	*Order out of Chaos: Man's New Dialogue with Nature (1984)*	Complex systems as intrinsically irreversible mechanisms exerting force
JAMES GLEICK	*Chaos: Making a New Science (1987)*	How simple dynamical rules, often deterministic, can generate complicated and often unpredictable patterns in nature and culture
M. MITCHELL WALDROP	*Complexity: Science at the Edge of Order and Chaos (1989)*	Self-organizing systems of interacting strategic agents
MURRAY GELL-MANN	*The Quark and the Jaguar: Adventures in the Simple and the Complex (1994)*	Contrasting a universe dominated by symmetry with life, defined in terms of frozen accidents describing effective information
JOHN H. HOLLAND	*Hidden Order: How Adaptation Builds Complexity (1995)*	Nested models of agents and meta- agents built from seven putatively universal properties and mechanisms
ROGER HIGHFIELD AND PETER COVENEY	*Frontiers of Complexity: The Search for Order in a Chaotic World (1996)*	Emergence of functional states of order from collective dynamics—mechanisms moving in the opposite direction to reductionism

AUTHOR(S)	TITLE/YEAR	ABBREVIATED DEFINITION
HAROLD J. MOROWITZ	*The Emergence of Everything: How the World Became Complex (2002)*	A survey of 28 candidates of instances of emergence from the origin of the universe to technology and urbanization
STEPHEN WOLFRAM	*A New Kind of Science (2002)*	The complex world as an instantiation of many simple programs that can be understood in correspondence (equivalence) to formal languages executed on a digital computer
JOHN GRIBBIN	*Deep Simplicity: Bringing Order to Chaos and Complexity (2005)*	Rules of interaction and nonlinear dynamics underlying far-from-equilibrium systems, including biological life
MELANIE MITCHELL	*Complexity: A Guided Tour (2009)*	Networks of agents capable of information processing and collective adaptation through learning and evolution
NEIL JOHNSON	*Simply Complexity: A Clear Guide to Complexity Theory (2009)*	The study of phenomena emergent from a collection of interacting objects including in physics and biology
SCOTT E. PAGE	*Diversity and Complexity (2010)*	Emergent outcomes of diverse rule-following adaptive agents interacting in networks
MARK NEWMAN	*Networks: An Introduction (2010)*	Network structure as a defining feature of engineered and evolved phenomena
JOHN H. HOLLAND	*Complexity: A Very Short Introduction (2014)*	The emergent properties of adaptive populations of agents

AUTHOR(S)	TITLE/YEAR	ABBREVIATED DEFINITION
JOHN H. MILLER	*A Crude Look at the Whole: The Science of Complex Systems in Business, Life, and Society (2016)*	Mechanisms of construction utilizing noisy and nested feedback loops connecting distributed adaptive agents
GEOFFREY B. WEST	*Scale: The Universal Laws of Life, Growth, and Death in Organisms, Cities, and Companies (2018)*	The thermodynamics of living systems and their macroscopic consequences—species abundance, life spans, and urban scaling
STEFAN THURNER, RUDOLF HANEL, AND PETER KLIMEK	*Introduction to the Theory of Complex Systems (2018)*	The science of generalized matter interacting through algorithmic rule systems
JAMES LADYMAN AND KAROLINE WIESNER	*What Is a Complex System? (2020)*	Open systems with collective dynamics classified by mechanisms of emergence: including their numerosity, feedback, and diversity
HENRIK JELDTOFT JENSEN	*Complexity Science: The Study of Emergence*	All those sciences providing principles for explaining emergent phenomena (including purely physical and biological transformations)

2000 Ostrom (84 citations)
1999 Laughlin et al. (46)
1998 Amari (41)
     Bowles (193)
     Watts & Strogatz (27)
1997 West, Brown & Enquist (37)
     Wolpert (15)
1996 Gell-Mann & Lloyd (29)
     Odling-Smee, Laland & Feldman (40)
1994 Arthur (10)
     Crutchfield (84)
     Forrest et al. (12)
     Schuster et al. (32)
1993 Lansing & Kremer (22)
     Mitchell, Hraber & Crutchfield (34)
1991 Bialek et al. (22)
     Holland & Miller (11)
     Lindgren (21)
     Simon (23)
1990 Farmer (66)
     Wheeler (177)
1989 Arthur (35)
     Perelson (76)
1988 Eigen, McCaskill & Schuster (48)
1987 Bak, Tang & Wiesenfeld (6)
     Kauffman & Levin (35)
     Langton (25)
     Reynolds (41)
1986 Pearl (35)
1985 Prigogine & Nicolis (16)
1984 Wolfram (30)
1983 Feigenbaum (11)
1982 Dyson (12)
     Hopfield (32)
1981 Axelrod (58)
1980 Packard et al. (14)
1979 Parisi (13)
1977 Gould & Eldredge (98)
1976 Haken (3)
     Hasselmann (18)
1975 Sherrington & Kirkpatrick (11)

2000 1975 1950 1925 1900 1875 1850 1825

Citation Publication Date

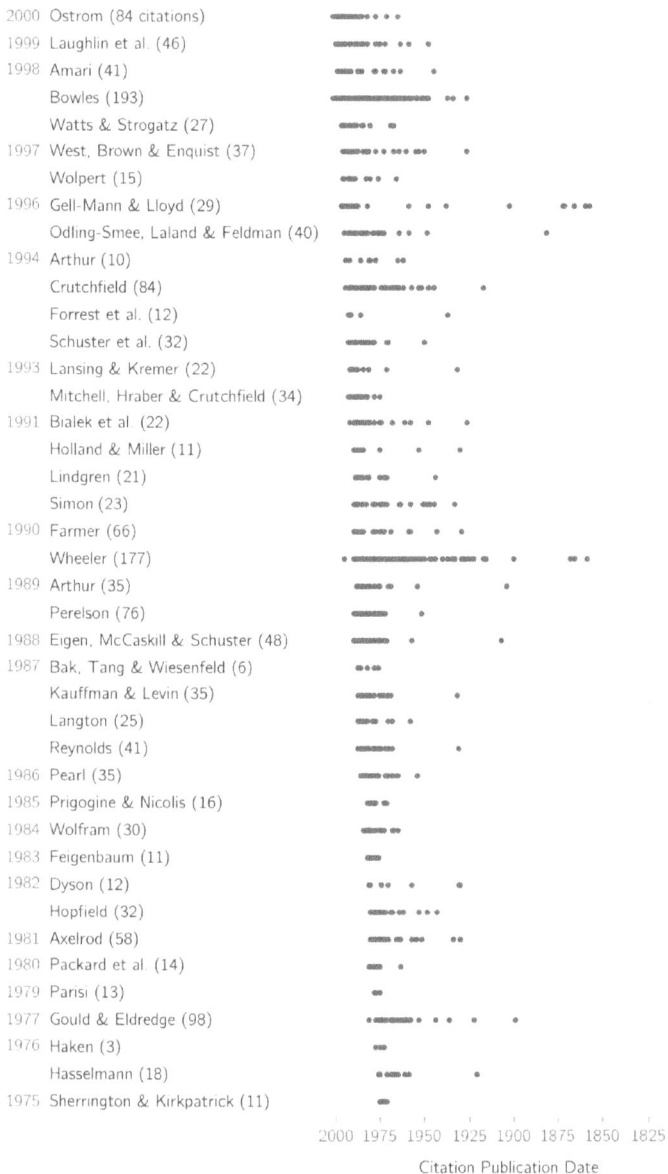

Figure 3. This visualization represents the date spans of works cited by the foundational papers that were published between 1800 and 2000 (only polymathic Wheeler 1989 went back further, to the 1600s). Each citation appears as a dot that corresponds to year of publication on the x-axis. The number of citations is indicated in parentheses.

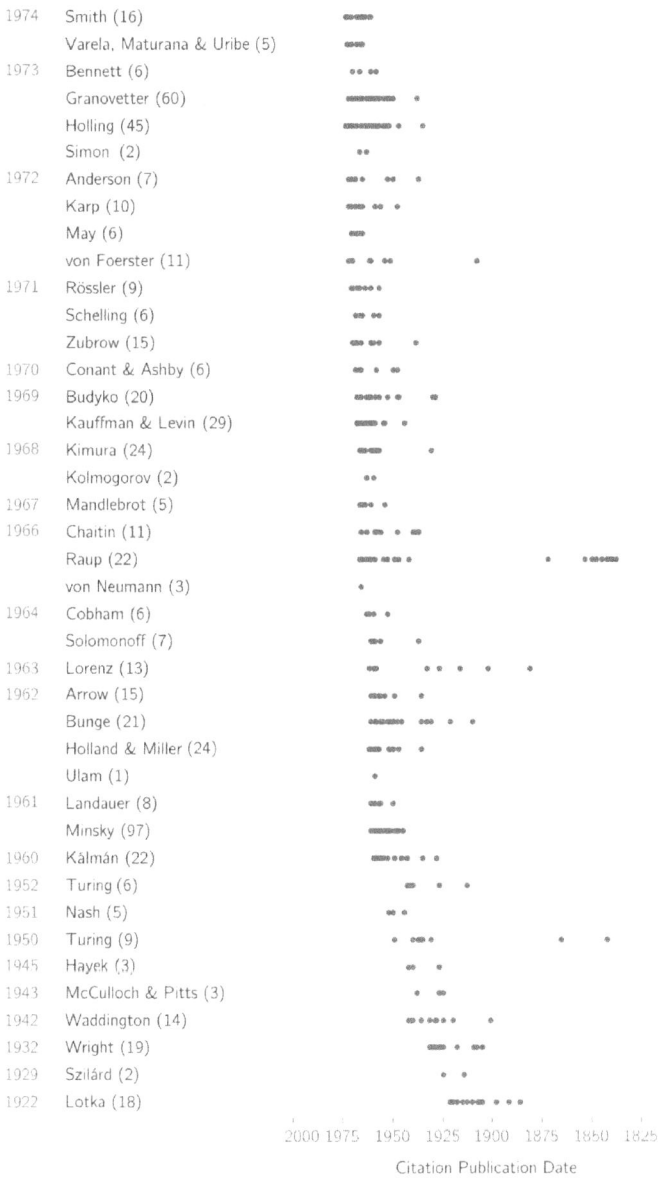

Year	Author (citations)
1974	Smith (16)
	Varela, Maturana & Uribe (5)
1973	Bennett (6)
	Granovetter (60)
	Holling (45)
	Simon (2)
1972	Anderson (7)
	Karp (10)
	May (6)
	von Foerster (11)
1971	Rössler (9)
	Schelling (6)
	Zubrow (15)
1970	Conant & Ashby (6)
1969	Budyko (20)
	Kauffman & Levin (29)
1968	Kimura (24)
	Kolmogorov (2)
1967	Mandlebrot (5)
1966	Chaitin (11)
	Raup (22)
	von Neumann (3)
1964	Cobham (6)
	Solomonoff (7)
1963	Lorenz (13)
1962	Arrow (15)
	Bunge (21)
	Holland & Miller (24)
	Ulam (1)
1961	Landauer (8)
	Minsky (97)
1960	Kálmán (22)
1952	Turing (6)
1951	Nash (5)
1950	Turing (9)
1945	Hayek (3)
1943	McCulloch & Pitts (3)
1942	Waddington (14)
1932	Wright (19)
1929	Szilárd (2)
1922	Lotka (18)

2000 1975 1950 1925 1900 1875 1850 1825

Citation Publication Date

Foundational papers that did not reference other works are excluded from this chart. *Examples:* Forrest *et al.* 1994 cites only a handful of recent publications, but Raup 1966 was significantly influenced by both work published in the twenty-five years prior and earlier publications from the first half of the nineteenth century.

Emergence and
Effective Theories

The prefoundational pillars, entropy, evolution, dynamics, and computation, all contend with questions related to the interface of local mechanisms with collective global properties. Maxwell, Clausius, Boltzmann, and Gibbs sought to map fundamental physical laws onto statistical regularities, or adduce justifications for a purely statistical theory of collectives. Darwin, Wallace, and Mendel explored the nature of persistent variation in populations, and how competition after sufficient time, produces an average macroscopic fit to the environment. Maxwell and Poincaré calculated how pairwise dynamical interactions might lead to long term stability or chaos at a system level. And Babbage, Lovelace, and Boole investigated how simple machines, implementing local logical operations, might support computational functions and even underpin patterns of thought.

We can with justification describe all four pillars as considerations of problems of emergence: the emergence of variables of state, the emergence of organisms and species, the emergence of fixed points and attractors, and the emergence of mind from machine. This is of course a very informal use of the word emergence, but nevertheless captures its essential concern: how to characterize and think about the origin of aggregate or "macroscopic" states of matter with apparently causal properties that are not physically, and "microscopically," fundamental.

The name given to theories that are not causally fundamental is *effective theories*. Effective theories discover and "label" large groupings ("coarse-graining") of fundamental or non-fundamental materials and properties that maintain coherence through time. These coarse-grainings are the collective, coordinated dimensions of matter and provide indirect evidence for emergence. More inclusive effective theories support a variety of models that exploit coupling among different effective theories achieving *inter-theoretic compatibility*. For example, the framework of population genetics connects two effective theories: (1) discrete, low-dimensional, genetic transmission (Mendelian genetics), and mutation-selection dynamics (Fisher–Wright model).

Consider for example how theories of condensed matter provide insights into the crystalline nature of metals. Engineers with limited knowledge of condensed-matter physics, including the condensed-matter physics of metallic bonds, or how material properties emerge from certain configurations of structured iron atoms and alloys, can still effectively shape metals into an engine block, and do so using abstract design principles.

At a far higher level of abstraction, we might be interested in the relative consistency and stability of psychologically and sociologically grounded affiliations within a polity: left or right leaning. These labels, and others like them, can serve as predictive factors for a wide range of seemingly independent behavioral and economic preferences, ignoring input from neural correlates and reported cognitive bias.

As long as variables assigned to each dimension maintain sufficient physical or organizational integrity through time, an effective theory has a chance of succeeding. In the final analysis, effective theories describe reduced dimensions along which coordinated aggregates of fundamental matter transform.

The more expansive problem of emergence is to explain both the origin and stability of these dimensions, and how these dimensions might influence subsequent emergence. Effective dimensions can be both many-to-one outputs of processes (e.g., atoms to metals; many neurons to one mind) and one-to-many inputs outcomes (e.g., educational attainment to income, mobility, and health).

Factor, Compression and Basis

Any model or theory confronted with non-fundamental observables needs to justify its choice of variables. More often than not, a weak form of justification consists in borrowing from the outlines of reality itself: clouds, cells, tissues, trees, species, cities, etc. This can of course be highly problematic since it relies on the stability of perception and description to isolate effective dimensions. More principled approaches abound. These include factor analysis, compression, and establishing the correct basis. None of these generate in themselves "effective theories," but they do provide useful pointers to where theories might apply.

FACTORS

In data analysis a very convenient approach is factor analysis. The basic principle is to relate a large number of observables, $(O_1, O_2, ..., O_n)$, to a smaller number of inferred factors, $(F_1, F_2, ..., F_m)$. If the observables are related linearly, then an obvious determination to make is how each hidden factor contributes (loadings) to each of the data points. If only the variance in the data needs to be fitted, *principal component analysis* (**PCA**) is a preferred method, where the right singular vectors from a *singular value decomposition* (**SVD**) are the effective dimensions of interest.

PCA provides some rather nice insights into a very limited, linear approach to data reduction and indirectly to emergence. It also illustrates the challenge of assigning the right labels to the inputs and outputs of emergence claims.

If every observable requires its own factor (they are independent variables), then the whole is equal to the sum of its parts. If the number of factors is small relative to the number of observables, then there is evidence of correlated behavior. These correlations shine a light on candidate effective dimensions.[6]

There is nothing in SVD–PCA ensuring that the factors are internal to a system. If the observables were measurements of some species' features, including abundance, metabolic rate, life span, and mobility rate, a perfectly credible candidate factor for every one is temperature. We might say that life span and mobility are the emergent outputs from the input temperature. This is nothing like water emerging from many H_2O molecules.

We also know that mass is a credible factor for these observables. Building an emergence explanation on mass is more familiar since it is a bulk property of the system itself. The way that mass influences each of these variables is provided by metabolic scaling theory: mass governs the efficiency of energy flows through fractal circulatory systems. In these situations the observables are emergent outputs from the input dimension mass.

COMPRESSION

Techniques of compression involve transforming data structures of a given size into smaller structures with a minimal loss of information. As with PCA, compression exploits the elimination of redundancies in data. Compression does not discover effective theories but suggests where they might be useful. Shannon–Fano codes and Huffman codes deploy codeword lengths in inverse proportion to word frequencies in order to achieve a lossless compression of a source file. Lempel–Ziv–Welch achieves lossless compression by building up a string-to-code lookup table as it receives a string, systematically replacing longer strings with shorter codewords.

_____

[6] Effective dimensions: The orthogonal dimensions of either a dataset or independent variables required by an effective theory.

The benefits of all forms of compression typically relate to the minimization of a resource requirement—space or time. This is somewhat analogous to replacing astronomical ephemerides with Newton's laws of motion, except in the case of a true law, the compression is asymptotically infinite as the dataset size increases.

BASIS

A different means of achieving compression is through a principled choice of *basis*. A basis is a set of vectors in a vector space that can encode every vector in the space. This is achieved using linear combinations from the basis.

The choice of basis tends to be domain specific: the spherical basis is useful in quantum mechanics; radial basis functions are commonly used to approximate partial differential equations; the Mexican Hat Wavelet is used to encode visual images. Some vector spaces are infinite or real-valued and involve infinite linear combinations. These include the periodic trigonometric functions combined into a Fourier series. These provide an orthonormal basis for the vector space of all real or complex valued functions.

Developing an effective theory is very closely related to the choice of basis; in fact it is typically dependent on the right choice of basis. The development of natural gradient optimization in neural networks was based on the field of information geometry (connecting probability theory to geometry by placing distributions on statistical manifolds). A natural metric on the manifold is the *Fisher metric*, a Riemannian metric, typically defined in terms of the exponential family of probability distributions. These distributions provide the basis for most of the results in the application of information geometry in natural science. The same could be said for the use of the Gaussian basis set in computational chemistry. These bases are commonly used to model the molecular orbitals (wave functions) of electrons.

Every effective theory has a basis and the basis provides either an explicit definition of dimension (e.g., Fourier basis) or some approximation thereof. Most commonly accepted examples of emergence, at least in relation to phase transitions, are expressed in terms of a basis. For example, Landau's theory of second-order phase transitions deploys a linear combination of the k-q function (quasimomentum and quasicoordinate space). Without the correct choice of basis, it is not at all clear whether an emergent phenomenon would be recognized.

More Is Different

For many scientists, the defining paper on emergence is Philip Anderson's "More Is Different" (1972). Its many profound ideas are lost when it is invoked to support a rather overworked sense of emergence, as understood by the Aristotelian aphorism "the whole is greater than the sum of its parts." It is in many ways a rather unfortunate adage. In a literal sense it has to be wrong, and in a conceptual sense it fails to convey why effective theories are important. Literally, a whole made from parts implies a reduction in number ($N \rightarrow 1$). Technically, an effective theory is always simpler, or more parsimonious, than the microscopic processes that it approximates. In both colloquial and technical usage, it would be more accurate to say, "the whole is less than the sum of its parts," in contrast to the expectation that it be "equal to the sum of its parts." This is what Newton had in mind when he wrote, "We are to admit no more causes of natural things than such as are both true and sufficient to explain their appearances" (Newton 1934, 398). What Aristotle and most successors have had in mind is better captured by the phrase, "different than the sum of its parts." It is this difference that Anderson explored in "More Is Different": the fundamental origin of epistemological irreducibility arising from scaling.

A single molecule of ammonia (NH_3) flips between two states—up and down—at 30 billion times a second. One consequence is that the stationary distribution of ammonia is a mixture of two mutually invertible pyramids. The barrier between two configurations of ammonia implies, as Anderson writes, that "the state of the system, if it is to be stationary, must always have the same symmetry as the laws of motion which govern it" (Anderson 1972, 394). When one considers larger molecules like phosphine (PH_3), the inversion rate is at least an order of magnitude slower. Phosphorus trifluoride (PF_3) is yet more massive and an order of magnitude slower than

PH_3. When one reaches biologically active molecules, at the scale of even the simplest carbohydrates, parity symmetry breaks down. The stationary distribution of molecules is dominated by initial conditions rather than fundamental laws of motion. The extra parameterization required in the initial conditions "counts" the broken symmetries, which stand as the foundation stones of emergence. *One of the neglected consequences of accumulating broken symmetries is that the most fundamental theories will be the least parsimonious.*

Anderson extends these insights in "Some General Thoughts about Broken Symmetry" (1994). Landau had written that "Every transition from a crystal to a liquid or to a crystal of a different symmetry is associated with the disappearance or appearance of some elements of symmetry" (Landau 1969, 25). For Anderson, the order parameter of a state of matter that changes through a phase transition should be thought of as a measure of the degree of broken symmetry, which in Landau theory is typically the degree of increased order. Since the order parameter is derived from an understanding of the symmetrical laws of motion, we get to see very precisely the degree to which the fundamental laws and emergent structures deviate from one another. This is why phase transitions are often mentioned in the same sentences as emergence: not because they are synonymous with emergence but because they provide evidence of symmetry breaking. Not always—the transition from a liquid to a gas is symmetry preserving.

In "More Is Different," Anderson presents a table with two columns, X and Y. Under X he includes a list of sciences that obey the laws of the sciences in column Y. For example, X (= chemistry) obeys Y (= many-body physics), and X (= cell biology) obeys Y (= molecular biology). The point is that "obey" and "determine" are not synonyms. We might say that the differences between obey and determine are statements about the number

of symmetries that have been broken. An operational definition of ontological reductionism is the limiting case where the
symmetries of the laws of physics determine the empirical distribution of observables. Under reductionism, the meaning of
obey and determine converge on identity.

In the study of complexity, effective theories that are deter-
minate for quantities of interest bear little resemblance to
the physics that they obey. At a certain point the symmetries
become so peripheral to a macroscopic behavior that even the
idea of broken symmetry loses its moorings, such that, "At some
point we have to stop talking about decreasing symmetry and
start calling it increasing complication. Thus, with increasing
complication at each stage, we go on up the hierarchy of the
sciences" (Anderson 1972, 396).

Mesoscopic Protectorates

Robert Laughlin, David Pines, and colleagues discuss a very compelling extension of the ideas of Anderson, particularly ideas of mesoscale organization, in "The Middle Way" (2000). Building up beyond broken symmetries, they generalize Anderson's analysis of ground states into *protected states*. These are states stable against small perturbations of the underlying equations of motion. Hence, "superfluidity, ferromagnetism, metallic conduction, hydrodynamics, and so forth are 'protected' properties of matter—generic behavior that is reliably the same one system to the next, regardless of details" (Laughlin et al. 2000, 32). Protectorates are the dimensions of effective theories. In the case of phosphorus trifloride, protection against tunneling is provided by properties of the chemical potential and the molecular wave-function. In the case of a large macromolecule like DNA, protection is achieved through a very complicated, evolved suite of error-correction mechanisms imposed on the primary information-transmitting structures. Emergence at living scales is perhaps best thought of in terms of mechanisms of protection, or robustness, against large perturbations enabling "effective" equations of motion "screening off" subjacent foundations.

The Republic of Effective Theories

In the natural and the social sciences, as well as in mathematics, nearly every theory is an effective theory: thermodynamics, condensed-matter physics, the theory of chemical reactions, population genetics, neural networks, neoclassical economics, game theory; theoretical computer science, set theory, and structural engineering. The domain of fundamental theory is highly restrictive and of little use either in theory-building or in practical applications. One might say to a first approximation that every model and theory deals with non-fundamental interactions and emergent properties.

Given this state of affairs, it is a little surprising that ideas of emergence seem so enigmatic. Part of the problem, as always, is the desire to find a single and simple model for a concept that describes a large family of processes.

The emergence of a liquid from a solid and the emergence of a mind from a brain have little in common materially. What they do have in common is a contingent set of effective theories (different ones) relating parts to wholes: in the first case, single molecules to incompressible fluids, and in the second, populations of cells to behaviors and perhaps even propositions of thought.

Practitioners are interested in the rich details of mechanisms, interactions, scaling, and limits. These details necessarily lie within a specialist domain, be it condensed matter, cognitive science, or psychology.

The value of framing dissimilar mechanisms as comparably emergent features derives from family resemblances. These are the structural and logical analogies observed across material systems. Through analogy we discover applications of frameworks and theories of great power far from their fields of origination. This includes the rather modest, albeit far-reaching, calculation of solutions to many-body problems (e.g., the use of the Ising model in physics, neuroscience, and social science), to the profound discovery of universal properties of collectives that transcend material differences (i.e., the universality classes).

Philosophies

The philosophy of emergence is largely concerned with the problem of relating fundamental theories to macroscopic features described through a suitable effective theory. This work has produced a number of fascinating concepts, including (Corradini and O'Connor 2010; Gibb 2019):

☛ *property dualism*—emergent states are produced by physical states but cannot be reduced to them.

☛ *coinstantiation*—mental events perfectly correlated with physical events can be treated as causal.

☛ *hierarchies*—nested effective theories, or degrees of coarse-graining.

☛ *transphysical laws*—or "emergent" laws describing persistent relationships between physical properties and secondary properties: from sensory experiences to metaphysical concepts.

☛ *multiple realizability*—many physical processes mapping onto a single aggregate process.

☛ *reductive unity*—a putative physical level underpinning all emergent phenomena.

☛ *conceptual entailment failure*—emergent phenomena that require different concepts to those used to explain parts.

Some of these concepts can be related to simple symmetry breaking during phase transitions—hierarchies, multiple realizability, conceptual entailment failure. Others can only be restrictively applied to function and minds—transphysical laws and property dualism.

Some philosophies are motivated by the puzzle that David Chalmers (2006) calls *strong emergence* or supervenience: how can effective theories be causal when there is only fundamental causality? There are areas of particle physics populated by those

Daniel Dennett describes as "greedy reductionists" (Dennett 1995, 82), only satisfied by explanations that miniaturize matter into particles and fields, achieving *reductive unity*. In the Appendix to Emergence, I discuss a possible resolution to this problem.

For most scholars, causality is almost always "effective causality," which makes it no less powerful. The steps in a mathematical proof follow from axioms in accordance with strict rules of deduction—that is, causal rules. The fact that these deductions are not described in the language of quarks and gluons has no bearing on the correctness of a proof.

The primary challenges of emergence are not concerned with relating the effective to the fundamental. This connection was even short-circuited in physics with Gibbs's revolutionary effective theory of thermodynamics; the most powerful and practical theory of thermodynamics to date. It is an approach to statistical physics that was supported by Maxwell's demonstration that intertheoretic reduction (from probability distributions to classical trajectories) is almost always impossible (logically and computationally). Greater attention is paid to inter-theoretic compatibility: Gibbs does not contradict Newton but cannot be derived from him. Life does not contradict quantum mechanics but cannot be derived from it. The pursuit of emergent descriptions is more often concerned with compatibility.

A confusion also arises from the way the word reductionism can be used as an antonym of emergence. It should be stated forcefully that there can be no theories of emergence without reductionism. Reductionism is used in two senses: reductionism as reducing all phenomena to a minimal set of elementary particles, waves, fields, symmetries and so forth (ontological); and reductionism as reducing a model and theory to an empirically justifiable minimal set of assumptions, transformations and interactions (epistemological). The first kind of reduction is only practiced by particle physicists and other fields that envy their style of demolition. The second kind of reduction, described as parsimony or compression, is pursued to varying degrees by all scholars, from cell biologists to historians, for the simple reason that there needs to be a basis for selecting a finite set of causal factors from an infinite set.

In figure 4, four of the contributors to thermodynamics and statistical mechanics are placed in an epistemological–ontological quadrant. Only Maxwell's theory can be classified as a fundamental theory operating at the microscopic level, where fundamental causality applies. Boltzmann's ergodic hypothesis is macroscopic and he desired that it be proved on a fundamental basis. Clausius's theory of disgregation is described as microscopic but had no basis in mechanics; it has the peculiar status of being an effective theory at the microscopic scale. The Gibbs theory of ensembles is purely statistical and operates at a macroscopic level deploying effective theories. Practicing physicists would not hesitate to call the Gibbs theory causal—fully understanding that it rests on the unproven, perhaps unprovable, limit of a fundamentally causal microscopic mechanics.

Figure 4. A quadrant of degrees of reduction. On one axis, ontological or spatial reduction, and on the other, epistemological or descriptive reduction. Even within statistical mechanics and thermodynamics, there are very different perspectives on what constitutes a fundamental and effective theory.

An Emergence Lattice

With these ideas in mind, and in order to establish analogical correspondences across different domains of emergence, this section introduces an *emergence lattice*. It is a simple schematic upon which different fields and disciplines can affix phenomena and theories. The purpose of the diagram is to establish the nature of the transformative processes that are being suggested as emergent and at what level in a hierarchy they apply.

The elements of the diagram can be enumerated:

☛ *Collections* (y-axis): the aggregation of individual units into collections of units

☛ *Transitions* (x-axis): the transformation of units or their collectives through suitable local or global mechanisms.

☛ Independent elements appear in lists surrounded by curly brackets: $\{x,y\}$.

☛ Elements conjoined ($\cup$) into stable units through transitions are surrounded by parentheses: (x,y).

☛ Collections indicate parts or units aggregated ($\in$) into multiplicities of parts or units: $n\{x,y\}$, $n(x,y)$.

☛ Entailment ($\leftrightarrow$) and failures of entailment ($\nleftrightarrow$) indicate intertheoretic reduction and non-reduction.

☛ The preservation ($\sim$) of molecular identity and abundance in different "phases": $n(xy)$ or $n(xy)'$.

☛ Different effective theories are indexed by their position on the lattice: $\mathrm{ET}_{(i,j)}$.

The first objective of the lattice is to clarify the very different senses of emergence implied by transitions and collections. In each domain of inquiry these map onto distinct mechanisms. *Transitions* (**T**) are typically local and pairwise. *Collections* (**C**) are typically global and involve many-body interactions. Secondly, whereas mechanisms of transitions are more often highly domain-specific, mechanisms of collections tend to be more generic (e.g., universality as captured through the renormalization group). Thirdly, different positions on the lattice do not all equate to different effective theories. Intertheoretic reduction, or entailment, can reduce the need for a new effective theory—no new theory is required to describe $n\{x,y\}$ not already contained within $\{x,y\}$.

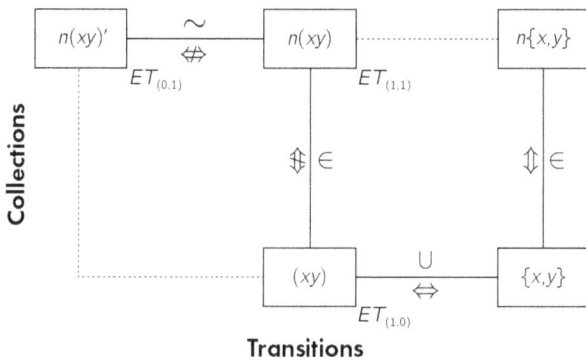

Figure 5. The emergence lattice is intended to keep track of emergence claims that are "horizontal"—Transitions derived from local or global interactions; and those that are "vertical"—Collections that move between individual and aggregate descriptions. Both possess system-specific and generic properties. Transitions are often described in terms of symmetry-breaking, Collections, in terms of coarse-graining and renormalization. Effective theory is abbreviated as *ET* in this and subsequent diagrams.

A Thermodynamic Lattice

The relationships between hydrogen and oxygen, individually and collectively, at different temperatures and pressures, provides a convenient example for the use of the emergence lattice. Enumerating each term individually:

☞ *Collections.* Molecular (H_2) and (O_2) exist as diatomic gases until sufficient energy initiates an exothermic reaction.

☞ *Transition.* The transition from molecular hydrogen and oxygen to a single water molecule involves the formation of three sigma bonds. These bonds are described using molecular orbital theory, which details constructive interference of the wave function of each atomic orbital.

☞ ($\leftrightarrow$). The structure of (H_2O) is entailed by knowledge of (H_2) and (O_2).

☞ ($\in$). The polarity of water molecules allows them to form hydrogen bonds in bulk, thereby aggregating into characteristic tetrahedral configurations in water.

☞ ($\leftrightarrow\!\!\!/$). The many properties of water, including ionic diffusivity, solvation, heat capacity, etc, cannot simply be derived from the properties of (H_2O) molecules.

☞ ($\sim$). At different temperatures and pressures, water undergoes phase transitions to solid and gas. The same Collections can exhibit different symmetries.

☞ (ET). Each phase has its own effective theory: Kinetic theory of gases, Navier–Stokes equations for incompressible liquids, theory of crystal lattices for solids.

Emergence as a Transition (e.g., solid to liquid) is not at all like emergence as a collection (molecule to water). When we speak of emergence in the context of the formation of water, we are using ambiguous shorthand for the theory of inelastic collisions (T); the theory of chemical bonds (T); the theory of phase transitions (T); the renormalization group (C); and collective currents (C). Each one of these theories relates to a large class of distinct mechanisms (of varying degrees of "fundamentalness") often highly dependent on the matter involved.

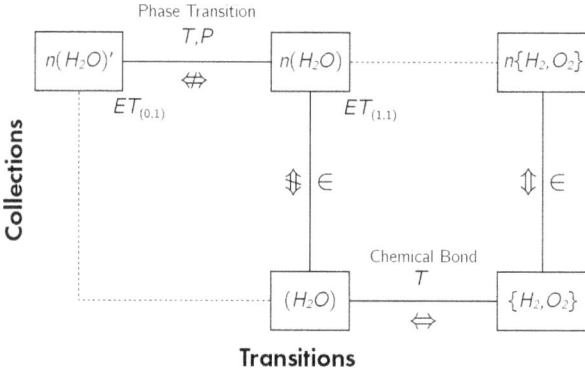

Figure 6. The Emergence Lattice of Thermodynamics. Temperature is abbreviated as T and pressure is abbreviated as P in this diagram.

An Evolutionary Lattice

A far more complicated case to be analyzed with the emergence lattice involves an evolving lineage. Entailment fails nearly everywhere, and effective theories proliferate. There are no canonical theories for transitions, and collectives involve so many heterogeneous interactions that it is impossible to predict detailed properties of the aggregates. In addition, each configuration is able to evolve an internal *schematic* encoding of its environment (genome, brain, etc.)—constituting an endogenous effective theory. In the following list, terms are placed together as they appear grouped on the lattice:

- ☞ *Collections.* (∪), (↮), (ET). Individual lineages form into an ecosystem defined by energy and mass transfers.

- ☞ *Transitions.* (↮), (ET). Lineages can fuse through symbiosis, forming a stable cellular unit.

- ☞ *Transitions.* (↮), (ET). Cells transition through genotypic or phenotypic modification.

- ☞ *Collection*, (∈), (↮), (ET). Cells cooperate into multicellular organisms with significant cellular division of labor.

- ☞ Evolution, (ET). Lineages of organisms do not transition directly one into the other. Transitions rely on subjacent transitions, followed by collections at the cellular level and below, in order to produce variation for competing populations at the multicellular organismal level. This selection process relies upon extraordinarily high-dimensional inputs from the environment (treated below as emergence in knowledge environment).

- ☞ (S). Unique to evolving lineages are internal schema (a simple internal effective theory of the world) required to persist and survive. These schemas encode sensory, memory, information processing, and behavioral capabilities. They are not constant over time.

The evolutionary example almost breaks the lattice. There are so many candidates for emergence and so few established mechanistic theories for how these might come about. Emergence refers to the origin of cells with cellular-properties, tissues and organisms with unique functions, the introduction of novelties through time over the course of evolution, and most importantly, an endogenous form of effective theory, a schema. The immediate consequence of a schema is that it blurs the boundary between the epistemological and the ontological. Every organism is both the object of a theory and a theorizer about the environment in which it lives. Human epistemology is the cognitive and cultural aggrandizement of this schematic feature of complex systems.

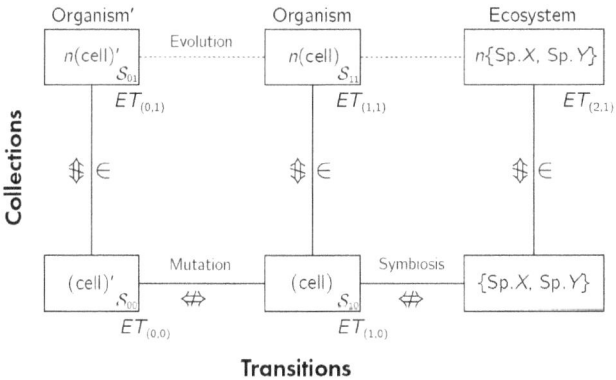

Figure 7. The Emergence Lattice of Evolution. Schema is abbreviated as S in this diagram.

Knowledge, Transitions, and Collections

Directly comparing the thermodynamic lattice and the evolutionary lattice obscures one of the most important considerations of complexity science: the disparity between a small number of global control parameters in a purely physical setting and a vast number of local selection pressures in the adaptive setting.

Darwin described the environment as an almost infinitely discerning "observer" of biotic variability—"an entangled bank." (Darwin 1859, 489) In order to facilitate a transition between any two lineages or species, an unknown, very large number of internal degrees of freedom must be modified; these form the basis of each genome, agent, or organismal schema. This is not a context where the theory of phase transitions has much to offer about biological emergence.

Let's call the typical physics theory "knowledge second" or **K2**, and the complex situation "knowledge first" or **K1**. A **K2** system produces knowledge from very limited inputs (knowledge from limited data). In physical and chemical theory, these are the laws of physics, initial conditions, and suitable control parameters. Using these minimal assumptions we are able to provide models and explanations for a wide range of observations from quantum mechanics to condensed matter and celestial mechanics. Eugene Wigner (1964, 995) describes physics in the following terms: "The elements of the behavior which are not specified by the laws of nature are called initial conditions. These, then, together with the laws of nature, specify the behavior as far as it can be specified at all: if a further specification were possible, this specification would be considered as an added initial condition."

A **K1** system produces knowledge from a vast number of inputs (knowledge from nearly unlimited data). In biological theory these are described as the environmental selection pressures acting on lineages of organisms; in psychology the childhood environment of assimilation; and in anthropology the facts of culture. Unlike initial conditions and simple macroscopic variables, including temperature and pressure, selective inputs are able to fine-tune through feedback high-resolution states of everything from DNA molecules to the form and function of tissues, and the "circuits" of the brain. These are the schema (S) of complex matter.

Ideas related to emergence, or how scale affects phenomenology, sit very naturally in **K2**. This is because the objective of **K2** theories is to demonstrate how parsimonious mechanisms applied to large collections of identical, or similar, elements produce outputs that require much scientific knowledge to explain. As described previously, the canonical example is how the collective properties of vast numbers of identical H_2O molecules produce a new phase of matter (liquid), by changing a single global input (e.g., temperature). This produces a state of order that is explained in terms of new forms of knowledge—effective theories—including the ideas of compressibility, boiling point, viscosity, solvency, and solubility.

When it comes to **K1** systems, the status of emergence is more ambiguous. This is because the ratio of inputs to knowledge outputs is large. A child learning a second language in a classroom is learning a preexisting knowledge structure. In fact, the language that any one individual holds in their head is self-evidently a tiny fraction of the total "dictionary" language available. It is also a small sample of the language to which any one individual has been exposed over the course of their education.

Many effective theories in complex systems are of type **K1** and are therefore theories stressing the environmental inputs: natural selection, linguistics, ecology, archaeology, economics, etc.

It is for this reason that biologists and social scientists use the words adaptation, learning, and evolution far more frequently than they use the word emergence.

An important challenge for complexity science is not only to explain how a small theory can account for considerable variability across a diverse range of physical phenomena (**K2,** e.g., Newton's theory of gravity), but how a small algorithmic theory (**K1**, natural selection, reinforcement learning, etc.) explains how vast numbers of input parameters produce a diverse range of adaptive phenomena—the difference between Ockham's razor, which applies to **K2** and meta–Ockham's razor (**K1**) (Krakauer 2023).

The **K1–K2** distinction lies at the heart of the emergentist attack on selection from biophysics and self-organization, and the more recent attack on intelligence in Large Language Models. Starting with D'Arcy Wentworth Thompson in his magnum opus, *On Growth and Form*, and continuing through the works of Alan Turing, Nicolas Rashevsky, Hermann Haken, Ilya Prigogine, John Conway, and Stuart Kauffman, there has been a concerted project to transform **K2** systems into **K1** systems. This has the character of substituting out a large number of local selection pressures or learning parameters (input contingent parameters) with a far smaller number of global control parameters.

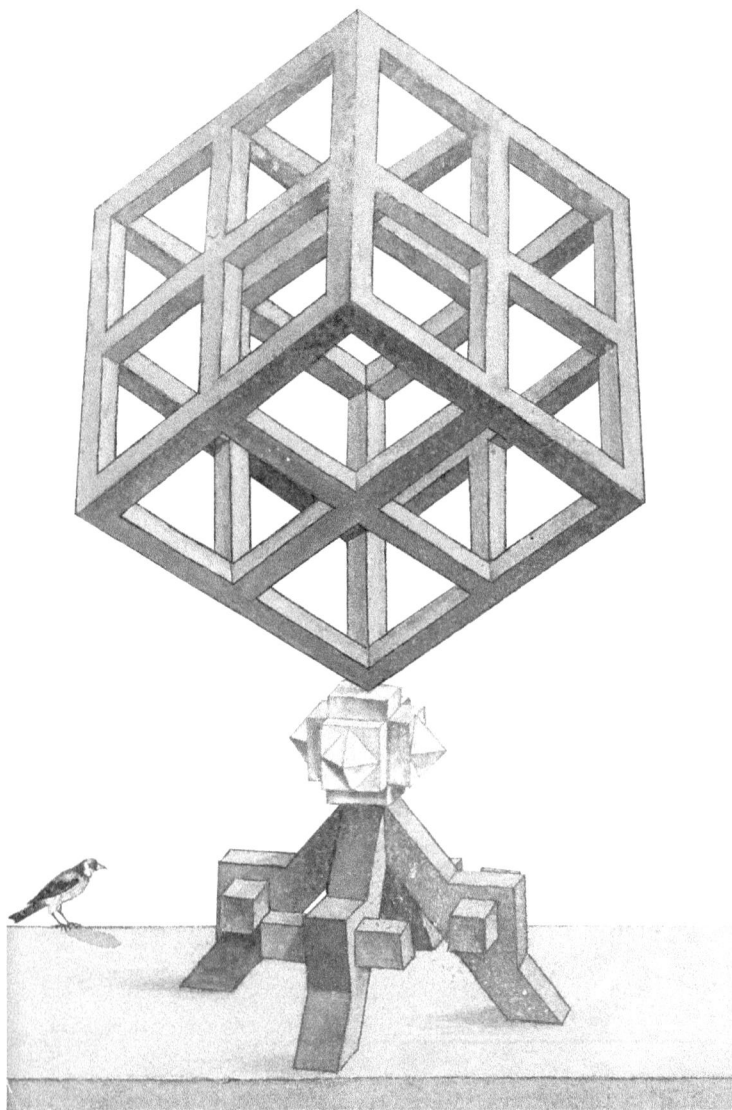

Information, Computation, and Cognition

T he development of complexity science is intertwined with the development of information theory, computing, computer science and engineering as they relate to information processing—cognition (Pylyshyn 1984; Piccinini and Scarantino 2011).

This is partly instrumental. The absence of closed-form solutions for many nonlinear dynamical systems, the use of stochastic simulations, the explicit use of computational concepts in fundamental scientific theories, and the purposeful or teleonomic behavior of agents induced an important mutual dependence between new complexity-related concepts and new information-processing frameworks. These include the consideration of computing resources and a variety of algorithms and heuristics. It is also partly fundamental—the complex domain as one where general information-processing principles apply at many scales.

In this section there is focus on a computationally oriented selection of the *Foundational Papers*: these describe a variety of proto-computational, computational, and cognitive systems (computing system, algorithms, and concepts).

Computational concepts began to be used in the 1920s and increased in use dramatically into the 1940s and 1950s, as physical computers became less scarce and computational principles became more important (Ceruzzi 2003). These principles include ideas relating to thermodynamics, calculation, filtering, sorting, controllers, Turing machines, transistor-like switches, and digital memories. By the 1960s there is scarcely a paper that does not rely on either computational hardware and software, and more importantly, place computational (information-processing) principles on an equal footing with those of physics, chemistry, and biology (Zenil 2013). This section is not about the "use of computers" in complexity research, or the extraction/discovery of patterns in complex systems through simulation, but a consideration of the diversity of computational principles and their sequelae as they *inhere* in complex systems.

Table 4 is organized along similar lines to David Marr's 1982 framework for levels of information-processing system (Marr 2010). The "Computing System" column describes the "category" of logical-machine upon which processing is assumed to take place; physical (as implied in the original Marr scheme) or abstract (e.g., cellular automata on a digital computer). The "Algorithm/Rule" describes the local operations acting on states of the machine. "Computational Concepts and Functions" are the desired outputs, their properties, and the principles of the computation. "Research Hardware" (when relevant and where available) describes the particular devices used by researchers in production of the publication when essential for the findings.

Conceptual Material

There is something wanting in the usual description of computation as the "execution sequences of halting Turing machines (or their equivalents)," just as there is limited insight derived from defining cognition in terms of "acquiring knowledge and understanding through thought, experience, and the senses." Both definitions clearly refer to a vast subterranean knowledge base. And yet anything we might care to describe or model in nature requires that we operationalize Turing machines, understanding, and thought.

In an important clarification for both computation and cognition, Edsger Dijkstra distinguished computation—the trajectory of a dynamical system—from an algorithm—the logical specification of a procedure guiding the computation. Algorithms are like recipes and computation is like cooking. The programmer, by applying an algorithm, structures "what is happening where in a useful way."

A decade later David Marr made a similar distinction when describing the visual system. Marr thought of computation in terms of processing of sensory inputs, algorithms as the specifications for how this processing should proceed, and the implementation layer described in terms of neurons, their properties, and their connections.

Both Marr and Dijkstra proposed that a science of computation or cognition should not be confounded with physical material and artifact but understood in terms of *conceptual materials*. This suggestion shuffles around the subject-object boundaries established in the physical sciences where matter moves and minds cogitate. A fairly representative take on the role of traditional physical theory is described by Roger Penrose (2004, 1,024):

> *Nevertheless, I should make it absolutely clear that the apparent lack of objectivity is not the fault of Nature herself. There is an objective physical world out there, and physicists correctly regard it as their job to find out its nature and to understand its behaviour.*

Which is as it should be. But when a natural structure is a purposeful mechanism, encoding both history and policy, and adapting to its own environment, then a *Dijkstra–Marr decomposition* (**DMD**) might be a better fit. It is rather revealing to look at a number of the Foundational Papers through the lens of the DMD. These illustrate how divergent studies in complexity can be from the standard model of a material or physical reality (ontology) cleanly separated from the logical methods of their analysis (epistemology).

In the following sections, the insights of the contributors are recruited to explain key ideas in the development of computation and cognition applied to complexity.

1920–1930

DEMONS: SZILÁRD (1929)[7]

The first study to take a DMD-like approach did so in order to resolve the challenge to the second law made by James Clerk Maxwell—the problem of "Maxwell's demon" as described by Lord Kelvin (Thomson 1874). Leo Szilárd in his paper "On the Decrease of Entropy in a Thermodynamic System by the Intervention of Intelligent Beings" (1929) opted to put the ghost back into the machine. The demon's intelligence consists in measuring and remembering microscopic states of a stochastic process. The curious thing about the Szilárd approach, building on Maxwell, is that it performs computational thought experiments at a fundamental scale. Susanne Still notes how "Szilárd's work . . . plays a role in the recent developments in far-from-equilibrium (also called "stochastic") thermodynamics, in situations in which the protocol applied to a thermodynamic system is dependent on measurements of (some aspects of) the system's state. . . ."

---

[7] This and subsequent case studies refer to articles featured in *Foundational Papers in Complexity Science*, e.g., volume 1, chapter 2. See pp. 184–191 for the contents of all volumes.

1930–1950

NEURONS: MCCULLOCH AND PITTS (1943)

Warren McCulloch and Walter Pitts (1943) present the first successful effort to connect a physical entity—the neuron—to both algorithms (logical circuits) and functions (logical propositions). A brain is every bit as physical as a galaxy, but is insufficiently understood by any physical effective theory. The brain needs to be understood by a qualitatively different class of effective theories, those we describe as cognitive or computational. Bruno Olshausen and Christopher Hillar dwell on one of the more surprising contributions of this work: "Among the most enduring contributions of McCulloch and Pitts's paper are their hand-drawn diagrams of neural circuits hypothesized to perform the fundamental operations that would constitute a system of logical calculus. Today we readily recognize these diagrams as the logic circuits—for example, and-gates, or-gates, and not-gates—that lie at the core of every digital computer. However such circuits were not at all obvious at the time, and in fact had not yet been conceived." How delightful, and so at odds with much utilitarian thinking, that a project to connect the soft tissue of the brain with the lofty philosophy of the *Principia Mathematica* led to the logic circuit.

FEEDBACK: ROSENBLUETH, WIENER, AND BIGELOW (1943)

Arturo Rosenblueth, Norbert Wiener, and Julian Bigelow (1943) "erased the difference between machines, animals, and indeed, humans," as Andrew Pickering describes their contribution. The strategy they pursued was to abstract an object from its environment and analyze the relation of its outputs to its input in terms of an error minimization (function). And this was to be achieved through negative feedback (algorithm). No longer a computational description of a natural system per McCulloch and Pitts, but a project to engineer the ultimate tracking device. The cybernetic movement was an impulse that over time morphed into the current interest in agency. And it is perhaps surprising that, as Pickering sees it, "Cybernetics is much discussed these days in the arts and humanities, and less so in engineering, but echoes of 'Behavior, Purpose and Teleology' are heard all over the disciplinary map."

INFORMATION: SHANNON (1948)

Working at Bell Labs, Claude Shannon wrestled with the practical computational challenge of transmitting a message reliably across space. This requires an algorithmic means of encoding a message with sufficient redundancy to be able to error-correct. The surprising thing about information theory is how widely applicable it has become, and how little this seems to depend on the underlying physical system. Seth Lloyd suggests that Information theory has in part been successful because "[t]he mathematical techniques that Shannon introduces are powerful but subtle. One of the triumphs of 'A Mathematical Theory of Communication' is that it uses multiple examples to present these techniques in a manner accessible to a broad audience." There is perhaps a lesson to be learned from Shannon's felicitous presentation of demanding ideas, which no doubt contributes to the consensus view that, "information and information processing lie at the heart of the sciences of complexity."

1950–1960

INTELLIGENCE: TURING (1950)

By the time Alan Turing wrote his paper, "Computing Machinery and Intelligence" (1950), he had jettisoned all consideration of the physics and chemistry of matter in favor of a purely functionalist approach. It is like Marr without the implementation layer (one might say he reads like Marr without a brain). Rather than design an experiment to probe the laws governing matter, Turing invented a new kind of experiment—"The Imitation Game"—to determine whether a computer is intelligent. It is hard to overstate how radical this move was: away from the methodology of particle accelerators, towards the analyst's couch. The many and various implications of this new and distinctly positivist verification principle are explored by Daniel Dennett through "what I will call anthroponormative thinking about thinking: our way of thinking, the human way of thinking, is the only kind of thinking worth considering."

1960–1970

STATES: KÁLMÁN (1960)

Several of the logical extensions of cybernetics to include multi-input–output, non-stationary processes, and internal states, were made by Rudolf Kálmán in the 1960s by means of the concepts of controllability and observability. Maxim Raginsky argues that "Kálmán made it possible to speak in a unified way to the interplay between structure, organization, and function of open dynamical systems." Kálmán augments the feedback narrative of Wiener and colleagues by integrating in ideas from nonlinear dynamics and thermodynamics.

BOUNDS: LANDAUER (1961)

In 1961 Rolf Landauer reintroduced physics to computation. Reestablishing the connection that had begun with Szilárd and that Turing, following a purely logical approach, had severed. In the creation of the field of the "thermodynamics of computation," Landauer demonstrated how any elementary computation implied a minimum energetic costs ($kT \ln 2$), now described as the minimal entropy flow. As David Wolpert observes, in order to arrive at an understanding of the physics of computation, a "revolution in physics" was required, and this in turn has provided some of the theories to "fully and formally analyze the thermodynamics of what we now call complex systems," Thus a computational approach along the lines of DMD loops back down to change the way physicists conceive of reality.

MINDS: MINSKY (1961)

In the same year, 1961, Marvin Minsky pursued the functionalists' approach to its limits: replacing algorithms with approximate heuristics and, like Turing, paying no heed to physics, chemistry, or biology. For Minsky, intelligence—which he thought of as generalized problem solving—was largely a matter of solving problems through "incomplete analysis." As Melanie Mitchell explains, Minsky's deliberate disavowal of the work of McCulloch and Pitts was one of its failings, and many today are "proposing that true machine intelligence will require deep neural networks to be integrated with . . . symbolic processing."

ADAPTATION: HOLLAND (1962)

Despite the obvious evolutionary foundations of complex systems, it was not until John Holland's paper in 1962 that a serious attempt was made to integrate population genetics with the theory of computation. Rather like McCulloch and Pitts seeking to reconcile mind-level-logic with a cell-level-calculus, Holland wanted to re-describe adaptation as a logical procedure: adaptation as a large distributed natural algorithm. Holland's approach, notes John Miller, was to imagine "a machine that is able to generate any possible program that could be run on Turing's universal computer. Adaptation is viewed as a process that modifies such generation process in response to feedback from the environment . . ."

PATTERNS: ULAM (1962)

It is natural to relate computation and cognition to questions of intelligence. One might just as effectively consider computation as an alternative framework for deriving the rules governing complex reality more generally. This was the perspective articulated in 1962 by Stanisław Ulam. Rather than "solve problems" à la Turing and Minsky, Ulam sought to generate patterns. Erica Jen explains that "The objective was to determine global characteristics such that limiting density of cells and to describe the dynamics . . . of the coherent structures generated in the spatiotemporal evolutions." By inventing cellular automata and exploring their behavior through exquisite visualizations, Ulam established one of complexity science's credos—unexpected forms of order arising through the collective iteration of simple, discrete rules.

CHAOS: LORENZ (1963)

There are few words as closely allied to complexity as chaos. Whereas complexity encompasses a very large number of phenomena, chaos is more restrictive in both derivation and application. And yet chaos strikes at the very heart of all mechanisms, exposing the ultimate limitations on prediction, and thereby erodes one of the dominant criteria used to evaluate the correctness of a theory. Chaos was discovered through computation, but, more profoundly, it reveals the limitations of any analog computational system. Given that the natural world is analog, and only approximates digital behavior, the reach of chaos is large. Doyne Farmer writes that "without chaos, the world would be extraordinarily boring: life and thought would be impossible without it."

DIFFICULTY: COBHAM (1965)

The concept of time is a fascinating conundrum in physics, but it is an existential challenge for complex systems. Questions relating to estimated life span, developmental rate, transient states of order, evolutionary fixation, and solution time, are all apposite for understanding a complex system. The problem of understanding the time-scaling of algorithms (how solution time scales with problem size) has emerged as one of the more rigorous definitions for computational complexity. Alan Cobham's paper, "The Intrinsic Computational Difficulty of Functions" (1965), provides an insight into what is meant when we state that a solution will take a long time to calculate independently of the physical computer upon which it is realized. Cris Moore writes that "Cobham was among the first to argue that problems have an intrinsic complexity—an objective fact about their structure, not a subjective one about whether or not we have found a clever algorithm." Cobham provided a principled insight into those problems for which the choice of physics is merely a constant of proportionality.

INDUCTION: SOLOMONOFF (1964)

A question related to intrinsic difficulty is how much contingent information from a process is required in order to identify its generative model—the problem of induction. Evolution through natural selection samples a small number of environments in order to produce adaptations of general value; learning involves experiencing a small training set in order to establish rules for unfamiliar contexts. In both evolution and learning, traits and knowledge are acquired that enable subsequent induction. Ray Solomonoff tackled these problems in his "Formal Theory of Inductive Inference" (1964). As Paul Vitányi describes it, "Solomonoff's ultimate goal in AI, as pursued in many subsequent papers, was a general system for machine learning. The goal is not so much to acquire knowledge itself but rather to determine how learning is performed by machines." Solomonoff's solution was to build a hierarchy of functions in which new problems access lower levels from earlier inputs. Unlike Turing, who presented a deductive framework with a one-dimensional architecture, Solomonoff arrives at a modular architecture for induction comparable to those described by Warren Weaver and Herbert Simon as hallmarks of complex systems.

COMPRESSION: CHAITIN (1966)

Whereas Solomonoff sought to induce general rules from short sequences, Gregory Chaitin sought to find the shortest programs to encode deductive systems. It seems somewhat intuitive that the length of a program to solve a problem should be related to the difficulty of the problem, following a logic along the same lines as Cobham's investigations. If one were to build a Turing machine, would there be a minimal way to do so, and how would this scale with the challenge? Simon DeDeo describes Gregory Chaitin's paper, "On the Length of Programs For Computing Finite Binary Sequences" (1966), as one that "twists back and forth between the mind-numbingly tedious pretense that a Turing machine is something one might actually want to construct in a materially efficient fashion (e.g., the endless design specifications of section 1.3, and the excursion of section 2, which feel like a user manual for a science-fictional device), and the mathematically mysterious idea that some sequences of binary digits might be more difficult—as a matter of engineering—to calculate than others." Chaitin's idea of complexity can be mapped in approximate language to genomes, neural schema, and machine weight vectors. To the extent that these have been minimized, they tell us something about the complexity of the problems that they are solving.

REPLICATION: VON NEUMANN (1966)

There is no evidence that Dijkstra and Marr ever met. But one might be forgiven for believing that they met virtually—in the imagination of John von Neumann. In an effort to understand the elaborate architecture of the brain, and of life itself, von Neumann chased down the quantitative implications of a number of computational metaphors and models. Von Neumann demonstrated considerable wisdom when he wrote "Now, none of this can get out of the realm of vague statement until one has defined the concept of complication correctly. And one cannot define the concept of complication correctly until one has seen in greater detail some critical examples, that is, some of the constructs which exhibit the critical and paradoxical properties of complication." Despite this assertion, von Neumann left a great deal in his writings ambiguous, but this barely diminished their influence. As Neil Gershenfeld describes the research: "Unusually, given von Neumann's mathematical rigor, he introduced a key concept of 'complication' without defining it beyond an ability 'to do things.' He doesn't quite mean complexity; he's getting at the minimum requirements for a system to reproduce and adapt. His description of complexity anticipates important results to come, including the work he presented in lectures at Caltech in 1952 on computing reliably with unreliable devices and the recognition of the physical limits of computing."

COMPLEXITY: KOLMOGOROV (1968)

Much as Chaitin builds on Turing to explain complexity, Andrey Kolmogorov arrives at similar conclusions by building on Shannon and the theory of partially recursive functions. Simon DeDeo explains the idea in terms of "'[s]imple' objects, ones with little algorithmic information, [which] have succinct programs that generate them; the Mandelbrot set may look complicated, but only a short program is needed to cover a wall with its patterns. 'Complex,' 'information-bearing' objects, meanwhile, have only long programs, full of exception cases and details." Kolmogorov moves in the direction of restoring to communication what Shannon very deliberately left out—semantics and behavior. Whereas information is rather like reporting only the caloric content of a recipe, Kolmogorov complexity tells us something about its ingredients. This is a computational step closer to what we mean to convey by "complex."

Table 4. A chronology of foundational papers exploring the interface of complexity and computation. Each contribution is described in terms of (1) the basic computing system or logical machines on which (2) algorithms or rule systems operate in order to (3) produce a desired function or output.

FOUNDATIONAL PAPER & COMPUTING SYSTEM	ALGORITHM/ RULE	COMPUTATIONAL CONCEPTS & FUNCTIONS	RESEARCH HARDWARE
SZILÁRD (1929): Physical systems and observing device	Measure/filter	"Intelligent" selecting mechanism with a memory	N/A
MCCULLOCH AND PITTS (1943): Physical neurons	Integrate and fire	Recurrent propositional calculus; "emergent parallel computation"	N/A
ROSENBLUETH, WIENER, AND BIGELOW (1943): N/A	Negative feedback	Servomechanisms: passive and predictive; (extrapolative) input–output behavior	N/A
SHANNON (1948): Telegraphy; vocoder; SIGSALY system	Discrete transducer	Optimal encoding and decoding of messages; information; error correction	N/A
TURING (1950): Universal machines	Store; execute; control	Universal machine; imitation game; artificial intelligence	N/A
KÁLMÁN (1960): N/A	Control grammarian; linear quadratic regulator	Controllability and observability	N/A
LANDAUER (1961): Magnetic films; cryotron; flip-flop circuits	Moving particle in bistable potential	Irreversible switches; minimum energy bounds; entropy production; entropy reduction	N/A

FOUNDATIONAL PAPER & COMPUTING SYSTEM	ALGORITHM/ RULE	COMPUTATIONAL CONCEPTS & FUNCTIONS	RESEARCH HARDWARE
MINSKY (1961): Turing machines; digital computers	Search; learning; planning; induction; pattern-matching	Artificial intelligence; problem-solving	N/A
HOLLAND (1962): Iterative circuit computer	Generators; generation trees	Logic of adaptation; universal generations procedures; populations of programs; adaptive computation	N/A
ULAM (1962): Cellular automata	Recursion relations; discrete dynamics	Pattern formation; growth and morphogenesis	IBM 704 36 bit; vacuum, tube
LORENZ (1963): N/A	Hydrodynamics: forced dissipative flow	Deterministic chaos; divergent trajectories; long-range unpredictability	Royal McBee LPG-20; 31 bit; vacuum, tube
SOLOMONOFF (1964): Turing machines	Formal grammars; stochastic processes; Bayes's theorem	Universal induction; algorithmic information theory	N/A
COBHAM (1965): N/A	Addition and multiplication	Computational complexity; polynomial versus exponential time	N/A

FOUNDATIONAL PAPER & COMPUTING SYSTEM	ALGORITHM/ RULE	COMPUTATIONAL CONCEPTS & FUNCTIONS	RESEARCH HARDWARE
VON NEUMANN (1966): Cellular automata	"Copy" and "paste"	Self-replications ("auto-reproduc-tion"); controller or "constructor" operations	N/A
CHAITIN (1966): Turing machines	Bounded-transfer Turing machine	Effective comput-ability; randomness and incompress-ibility	N/A
KOLMOGOROV (1968): Theory of recursive functions	Asymptotically optimal programs	Kolmogorov complexity; computability and randomness	N/A
CONANT AND ASHBY (1970): Morphisms of Turing machines	Regulation by error-control	Theory of mod-el-based regulation	N/A
SCHELLING (1971): Cellular automata, agent-based model	Discrete dy-namical updating in Moore neighborhood	Spontaneous order; emergent sorting/ segregation	"Tabletop" experiments
KARP (1972): Turing machines	Reduction; language; recognition problems	NP-complete property; combinatorial complexity	N/A
VON FOERSTER (1972): Recursive relations	"Computations" on representa-tions and their relations	Self-reference; theory of the observer	N/A

FOUNDATIONAL PAPER & COMPUTING SYSTEM	ALGORITHM/ RULE	COMPUTATIONAL CONCEPTS & FUNCTIONS	RESEARCH HARDWARE
BENNETT (1973): Turing machine	Logically reversible Turing machine	Thermodynami- cal reversibility; operating close to thermodynamic equilibrium; effi- cient computation	N/A
VARELA, MATURANA, AND URIBE (1974): Cellular automata	Catalysis and linkage	Autopoiesis; self-production; self-distinction	N/A
HOPFIELD (1982): Recurrent neural networks	Hebbian learning	Content-address- able memory; generalization; error-correction	N/A
WOLFRAM (1984): Cellular automata	Finite state transitions in 1D	CA classes (1–4); dynamical and computational classes; limits to prediction; compu- tational irreduc- ibility	N/A
LANGTON (1986): Cellular automata	Finite state transitions in 2D; sparsity control parameter λ	Equivalence of operators and operands; emergent phase-diagrams; molecular logic as propagating periodic structure	Apollo Corporation DN600; Dual 6800
WHEELER (1989): Physical universe and observing device	Measurement of bits; self- referential deductive systems	It from bit: "Communication employed to estab- lish meaning"; most elementary physics (quantum) involve observer-partici- pant in a Shannon channel	N/A

FOUNDATIONAL PAPER & COMPUTING SYSTEM	ALGORITHM/ RULE	COMPUTATIONAL CONCEPTS & FUNCTIONS	RESEARCH HARDWARE
HOLLAND AND MILLER (1991): Agent-based model	Genetic algorithms; classifier systems	Artificial adaptive agents (AAA); emergent economic phenomena	N/A
MITCHELL, HRABER, AND CRUTCHFIELD (1993): Cellular automata	Genetic algorithms	Emergent computational capability; evolutionary "programming" of parallel computers	N/A
ARTHUR (1994): Agent-based model	Genetic algorithms; temporally fulfilled expectations	Co-evolutionary induction	N/A
CRUTCHFIELD (1994): Bernoulli Turing machine	Epsilon machine reconstruction	Causal states; statistical complexity; entropy rates; finitary stochastic hierarchy	N/A
FORREST, PERELSON, ALLEN, AND CHERUKURI (1994): Digital computer; SPARC OS; DOS	Negative selection algorithm	Computational immune systems; self–nonself in computer security	N/A
GELL-MANN AND LLOYD (1996): Turing machines	Algorithmic information content	Effective complexity; total information	N/A
AMARI (1998): Neural networks	Natural gradient learning	Blind-source separation/ deconvolution	N/A

1970–1980

REGULATION: CONANT AND ASHBY (1970)

Roger Conant and W. Ross Ashby pick up the threads of the cybernetic–control narrative from Arturo Rosenblueth, Norbert Wiener, Julian Bigelow, and Rudolf Kálmán. Rather than move down in DMD into the implementation, they move several levels up into algorithmic abstraction. Like Kálmán, their interests lie in opening the black box of behaviorism favored by Rosenblueth et al. Somewhere in this box are echoes of Solomonoff, Chaitin, and Kolmogorov. The challenge of regulation must relate in some fundamental way to the complexity of the regulated. James Crutchfield writes, "Perhaps stating the opposite helps: if the regulator knew nothing about the system—nothing about its accessible configurations and states—it could do no effective control. The regulator could perturb the overall system, but those changes would not be coordinated with the system's moment-by-moment condition. And so the regulator would help as much as it hindered reaching the goal." Through the regulatory lens, complex reality is like a hall of mirrors composed of a near-endless series of reflections.

SEGREGATION: SCHELLING (1971)

How do these several computational ruminations bear on the complex systems in which we swim? Thomas Schelling took it upon himself to answer this question in "Dynamic Models of Segregation" (1971). Schelling is in the lineage of McCulloch, Pitts, and Ulam searching for simple rules that might reproduce the outlines of complicated patterns; he channels the parsimonious philosophy of Chaitin and Kolmogorov, for whom the complexity of a model might have something to say about the complexity of the social phenomenon. It is also an attempt to place Hayek's theories into a computational language. Peyton Young considers that "Schelling's 1971 article represents a major departure from this way of thinking.

Indeed, it represents one of the clearest expressions of the notion of spontaneous order and how it can be applied to a contemporary matter of practical significance, namely, racial segregation." Through an *agent-based model* (**ABM**), Schelling is able to connect ideas related to collective dynamics with simple heuristics to derive an emergent pattern of enormous social consequence.

REDUCTION: KARP (1972)

There is little sense to the question, "How intrinsically difficult is it to synthesize ethanol from ethylene and water?" We have a perfectly adequate theory of chemical reactions and the competence of a chemist should play no part in the correct answer. But there are problems where such difficult questions are natural, and these relate to calculating solutions to combinatorial problems. Some of the hardest of these problems are those that take an exponentially long time to discover but only polynomial time to verify: NP-hard. These problems might sound rather exotic, but in 1972 Richard Karp showed just how widespread NP-hard problems really are. As Cris Moore describes the situation, "We can then build simulations on top of simulations, creating a family tree of NP-complete problems. . . . This family tree has since grown to include hundreds of problems in algebra, automata, machine learning, statistical physics, calculus, and virtually any other system in which complexity can live." Once again, it is not the materials that pose the problem but the internal structure of a system—a property that yields to an algorithmic and computational description.

SUBJECTIVITY: VON FOERSTER (1972)

One of the consequences of the cybernetic movement was to alert the world to the ubiquity of feedback. Kálmán and Conant and Ashby presented some of these implications in terms of control and regulation. It was Heinz von Foerster (1972) who extrapolated these ideas to the limit where the regulator and the regulated coalesce into self awareness: "Abandon all hope, ye who enter here." Manfred Laubichler summarizes the dilemma: "Von Foerster's starting point is as simple as it is profound. Insofar as the observer is now a crucial part of any scientific account, we need a description of the observer to be part of any science. And, as the only observers we know are living beings (so far limited to our planet), we need a biologically sound theory of these observers and their ability to make these observations and express them in a coherent form." The information-processing implications of this situation remain an outstanding challenge for complexity, and are only likely to get worse, as we continue to replicate ourselves imperfectly in machine-learning memory.

REVERSIBILITY: BENNETT (1973)

Szilárd vanquished perpetual motion machines of the second kind. Landauer calculated the minimal costs of running a computing machine. Charles Bennett demonstrated that if a computation was reversible these costs might be driven to zero. Jon Machta explains how "Bennett refuted the belief that computing requires logical irreversibility and heat dissipation by showing explicitly that it is possible to design a general purpose logically reversible computer. . . What Bennett showed by construction is that any computation that can be carried out by an irreversible one-tape Turing machine can also be carried out by a three-tape Turing machine that is logically reversible." Bennett's extraordinary insight was to suggest a way in which an algorithmic account of computational work might trump the physical laws mandated by mechanical work. This has not yet happened.

AUTOPOIESIS: MATURANA, VARELA, AND URIBE (1974)

Investigations into the origin of life are traditionally conducted in laboratories. Prebiotic conditions are replicated and small molecules are provided. Elementary forms of biosynthesis are construed as plausible models for the transition from an abiotic to a biotic planet. Humberto Maturana, Francisco Varela and Ricardo Uribe started somewhere else: with the computational principles that these chemistries were aiming to manifest. Randall Beer writes, "an autopoietic system is one organized as a network of processes that have the dual properties of self-production and self-distinction. In order to concretely illustrate the idea of auto- poiesis and with the hope of spurring the development of formal tools for its analysis, Varela, Maturana, and Uribe developed and simulated a minimal computer model of the central ideas. The model takes the form of an artificial chemistry playing out upon a rectangular grid." The final sentence is an echo of Ulam and von Neumann and the prequel to a growing class of models that seek to discover the minimal set of rules inherent in life within the reasonable confines of an expanded Go board.

1980–1990

MEMORY: HOPFIELD (1982)

The logical properties of small neural circuits had been established by McCulloch and Pitts. The brain is made of billions of neurons and this begs the question of what new principles of computation might operate at larger scales. Drawing on insights from the collective dynamics of condensed matter, John Hopfield sought computational correlates corresponding to coordinated macroscopic states. David Sherrington explains the connections between computational neuroscience, dynamical systems, information theory, and statistical mechanics: "He also employed a very simple formulation for information coding in the synaptic connections, philosophically encompassing Hebb's observations on synaptic evolution, and with asynchronous dynamics. Hopfield noted that, for symmetric synapses, these assumptions led to a Lyapunov minimization dynamics resulting in associative cooperative dynamics toward retrieval of patterns coded in Hebbian synapses. This opened the door to Gibbsian statistical physics analysis and simulation, of a type that had recently been employed for spin-glasses." To this day, a tension remains between the intuitive appeal of simple circuits whose causal flows can be described, and the emergent properties of large networks, whose coordinated states perform computational work.

SPACE: WOLFRAM (1984)

Neither small Boolean circuits nor large collective networks elucidate the role that spatial patterns might play in computation. Building on Ulam's and von Neumann's interest in cellular automata, Stephen Wolfram initiated a project to connect these ideas directly to the logical concerns of Turing, Solomonoff, Chaitin, and Kolmogorov. What emergent forms of computation might we expect from compact rule-systems placed into discrete spaces? Wolfram suggested a taxonomy of pattern-formation corresponding to distinct dynamical and computational complexity classes. According to Hector Zenil, "The study of the limit behavior of small computer programs can provide an early indication of their computational capabilities without having to prove universality, which is known to be extremely difficult for general cases. The proofs of von Neumann, Konrad Zuse, or Conway for universality took decades to develop. It has been shown that pervasive computational universality, as defined by Wolfram's complexity classes, is likely to be very common and is suggestive of spatial computation in nature."

SIMULATION: LANGTON (1986)

The research of von Neumann, Holland, Varela, Maturana, and Uribe established connections between principles of computation and principles of life: algorithms that achieve replication, error-correction, and adaptation. Chris Langton pursued the DMD down into biochemistry, seeking a dynamical logic in which computational substrates could mirror those of chemistry. Langton was more interested in the dynamics of computation than the sequential recipe defined by algorithms. Sara Walker describes the profound functionalist implications of Langton's work: "If all the functions of life can be simulated in something like a cellular automaton, we can come to understand life purely by its logical and informational properties. We don't need to build life to understand it, we only need to simulate it." The debate over the adequacy of simulation is ongoing, ultimately turning on the question of whether the most causally significant factors live in the materials or the conceptual materials.

1990–2000

BITS: WHEELER (1990)

It is always exciting to encounter an apostate. And when they are leading congregationalists it is even better. John Archibald Wheeler was by any measure a considerable force in theoretical physics: an important contributor to general relativity and the theory of gravity, to fundamental theories of quantum mechanics, and to quantum information. Wheeler's apostasy was to take a coarse-grained theory of information and claim it is more fundamental than the microscopic laws of physics. Wheeler called this inversion "it from bit." Jessica Flack captures Wheeler's heresy: "Wheeler's most intoxicating conclusion is that information precedes matter, rather than the other way around, as is typically the case in grand unified theories. Wheeler's story turns not on objects like strings but on information-theoretic entities—decoders or registrants—his so-called observer-participants or choice-makers, as I call them" The extraordinary implications of this thesis is that physics should be viewed as a computation acting on information: complexity is foundational and physics an epiphenomenon.

AGENTS: HOLLAND AND MILLER (1991)

James Clerk Maxwell was astounded by the identical nature of every electron. His hypothesis to explain this uniformity was to posit a divine factory tuned to infinite tolerance. Physical theory takes homogeneity at scale as an input in order to develop powerful collective theories. Life is otherwise—fully hetero-geneous and deeply history-dependent. This is not more evidence than in societies, and yet canonical economics models are often in thrall to the uniform simplicities of the physical universe. John Holland and John Miller pursued a computa-tional approach to economics that fully encompassed human diversity, developing agent-based models. As Rick Bookstaber writes, "It seems self-evident that people are different from one another. We each have different ways of looking at the world, different approaches for making decisions. People also change. We are tempered by our individual experience and by changes in our environment, so how we make decisions today might not be the way we do so tomorrow. Such is the world in which we reside. It is also a world in which the artificial adaptive agents described by Holland and Miller reside." The reluctance to adopt ABMs is a complicated issue, in part a question of sensitivity analysis, but perhaps to a greater degree a persistent preference for mathematical formalisms in closed form, and an instinctual distrust of the computationally undecidable.

λ: MITCHELL, HRABER, AND CRUTCHFIELD (1993)

One of the greatest scientific insights into the natural world has been the realization that different forms of matter can be different phases of the same matter: water vapor, liquid water, and ice. The selection of phase depends on the value of a suitable control parameter (e.g., temperature or pressure). Using *cellular automata* (**CA**), Christopher Langton sought out a means of finding a control parameter (λ) describing the phases of computational patterns in terms of Wolfram's four complexity classes. But he went one step further: just as physical matter can be tuned to the vicinity of a phase transition, Langton suggested that computational forms of life would live at the "edge of chaos"—the boundaries of the complexity classes. Melanie Mitchell, Peter Hraber, and James Crutchfield discovered that, rather than tune to a critical value, evolved CAs tend towards values associated with the performance of a particular task; they tend to avoid edges. Dave Ackley summarizes the controversy: "For me, this particular story is about a shift from theoretical physics to empirical computer science. The idea that persists throughout is that physics and life are fundamentally linked, somehow, via computation and programming.

The change is that later work focuses less on global properties to be derived or formally proven, and more on concrete examples to be implemented and studied. In this story, statistical physics as a formal framework gives way to 'emergent physics' observed in empirical results."

IMMUNITY: FORREST *ET AL.* (1994)

The immune system is for many the canonical complex system: autonomous, decentralized, adaptive, coordinated, feedback regulated, and communicative. If nature can be rationalized as computational, what about rethinking computation in terms of the complexity of the immune system—or at least some algorithmic abstractions of immunity? This is what Stephanie Forrest and Alan Perelson did in their "Self–Nonself Discrimination in a Computer" (1994). Anil Somayaji describes the three influential contributions of this work: "The first was its demonstration that computational models of the immune system can be used to create novel algorithms. . . . The second key influence was its presentation of the negative selection algorithm. . . . The third key contribution of this paper was the idea of self versus non-self as applied to computer systems, particularly computer security." Forrest and Perelson's paper is a pioneering example of the use of biological models for technology development. Until recently, this has been rather limited. With the growth of brain-inspired neural networks and their surprising and astronomical economic impact, research along these synthetic lines is likely to exponentiate.

ε: CRUTCHFIELD (1994)

Since Kálmán, Conant and Ashby, and von Foerster, there has been a concerted effort to derive some index of the computational complexity of an agent. The approach of Chaitin and Kolmogorov assumed that the problem is known and needs to be parsimoniously described; Solomonoff came closer with his theory of inductive inference but, by basing his ideas on Turing machines, came up against the problem of uncomputability. In "The Calculi of Emergence: Computation, Dynamics, and Induction" (1994), James Crutchfield sought to derive a measure for an observed time series by constructing a minimal stochastic generator for its statistics—the (ε)-machine. Peter Sloot, Rick Quax, and Mile Glu describe Crutchfield's approach: "This agent is tasked with describing the observed process as efficiently as possible, i.e., by constructing an internal model which uses a minimal number of states to predict the observed future of the process. The overarching idea is that the extent to which the process is deemed 'complex' is defined by the agent's autonomous deliberation. More specifically, it is defined through the evolution of the size of the minimal model that the agent can formulate." Through this Conant–Ashby-styled lens, an agent induces a model from the world and then deduces its minimum—it thereby achieves an inferential cycle of adaptivity.

EL FAROL: ARTHUR (1994)

Game theory is a powerful formalism for studying strategic decisions at equilibrium. Over the past couple of decades, a number of formalisms have been developed to deal with the computational and cognitive challenge attendant on finding these equilibria; these include behavioral economics, Bayesian games, and algorithmic game theory. Brian Arthur's "Inductive Reasoning and Bounded Rationality" (1994) was one of the first to sacrifice the analytic for the synthetic, and do so using ABMs. Willemien Kets describes the problem as follows: "Suppose going to a bar is only pleasant if not too many people show up. How do you decide whether to go? This 'bar problem' was inspired by the popular El Farol bar in Santa Fe. But, of course, this toy problem is an exemplar of a much more general class of problems: How do people divide a scarce resource among themselves? . . . If the system is not in equilibrium, then at least one of the players could gain by changing her action. For example, if few people are going to the bar one night, then at least one player would be better off if she attended rather than staying home. And even if people face this problem repeatedly (say, every Thursday night) it is far from clear the system would converge to an equilibrium." Whereas the analytic approach divorces ontology and epistemology, an ABM builds learning and memory into the theory. It is like a mathematical morphism where we use a computational model of the mind to study the mind in the wild.

REGULARITY: GELL-MANN AND LLOYD (1996)

Theoreticians are nearly all minimalists. Given a choice between a Watteau or LeWitt painting, it would have to be the latter's bands, lines, and cubes over the former's wigs, gardens, and picnics. And yet the natural world can only be compressed so far, and at a certain limit, features begin to be lost that are constitutive of phenomena. Deriving the information-theoretic limit that is "as simple as possible but no simpler" is what Murray Gell-Mann and Seth Lloyd (1996) described as "effective complexity." Miguel Fuentes writes, "The effective complexity could be defined in colloquial language as the length of the compressed description of its regularities." In other words, not of its randomness—the bane of information-theoretic measures of complexity. But of course also of theorizing: "The sum of the effective complexity and entropy is called total information. . . . it is all the information required to describe an entity (or phenomenon). The importance of having this form of description lies, from my point of view, in the fact that it is very close to the task of critical knowledge taking, providing a way to evaluate or compare possible descriptions of the same phenomenon." Since William of Ockham onwards, careful thought has been synonymous with frugal hypothesis, and effective complexity puts more bits on the bone of the idea.

GEOMETRY: AMARI (1998)

There is a natural relationship established between learning and geometry through gradient descent. The question is, which gradient? The Euclidean gradient is the most familiar but it might not be the "natural" choice. Shun'ichi Amari showed that the natural gradient depends on the Riemannian metric tensor of the parameter space. In other words, the landscape should tell us how to move across it. As Nihat Ay writes, "The main idea of the natural gradient method is that a particular parameterization serves as a coordinate system for learning and has no special meaning. One can parameterize the actual search space in many equivalent ways. The geometry that we should use for optimization of the search is the one that is naturally defined on that space . . . respecting the geometry of the search space as outlined in this introduction greatly improves previously proposed gradient learning algorithms." An outcome of Amari's search for the natural gradient was the development of the field of information geometry—the study of statistical manifolds—spaces where every coordinate is an hypothesis. This brings us full circle to where theories of physical spaces (parameters) can be related to inferential landscapes (hypotheses): materials and conceptual materials are integrated within a suitable variational framework.

Concluding Unscientific Postscript to Complex Fragments[8]

Complexity science has been evolving for the last two centuries, starting with science inspired by machines—manufactured and evolved—and slowly fusing into a new epistemology.

Complexity science is perhaps the first modern science to transcend disciplines. The phenomena that it investigates, and the models and theories it deploys, move fluidly across fields.

This fluidity reflects an underlying ontological watercourse best described by self-organizing and selective principles operating on broken symmetries, and not by the time and space symmetries that yield to parsimonious frameworks of least action.

Information, energy, and computation provide a principled way to talk about randomness and order and define a continuum upon which to place both natural processes and cultural creations.

The adaptive sciences—evolution, learning, control theory, and computation—provided ideas to make sense of this reality. And, over the last several decades, these have recombined to generate many of the most novel ideas in complexity science.

---

[8] *Concluding Unscientific Postscript to Philosophical Fragments* was Søren Kierkegaard's 1846 pseudonymous defense of a subjective approach to knowledge. In it he writes that "The subjective existing thinker is aware of the dialectic of communication." This is an apposite framing for the spirit of this final section."

The growing computational–cognitive perspective on complexity shows how abstract ideas need to be rooted in physical reality (thermodynamics of computation and cognition) and how physical reality might emerge from information ("it from bit").

With the origins of life, multicellular organisms, and systems of knowledge, physics and chemistry have become auxiliary scientific frameworks in support of a variety of mechanisms of function (teleology).

The integrated nature of complexity science aligns with the connected nature of the modern world. Complexity science will be essential to all future projects that aim to escape terminal planetary decline. 🐦•

Appendix to Emergence

Supervenience, Programs, and Compilers

This section draws on emergence and computation to explore the idea of strong emergence in a causally consistent fashion. The world of apparently top-down causal engineered systems is presented as a useful analogy.

Supervenient phenomena that one should like to understand all connect *low-information sources* to *high-information targets*. This of course requires that the target augment information from the source in a predictable way. A few candidates include how a bird listening to a sound (information source) is able to reproduce that sound with its syrinx (information target); how reading a musical score (information source) can through performance produce structured sound waves perceived as musical melodies (information target); how learning calculus (information source) can through industry put a rocket into orbit (information target); and how vast corpora of data (information source) enable a large language model to pass a Turing test (information target). In each of these examples, low-dimensional inputs—frequencies or symbols derived from a finite alphabet—influence the form and function of very high-dimensional coordinated physical matter at a fundamental level.

The concepts that we shall use to explore supervenience are derived from computer science: programming languages (Sammet 1969; Davis, Sigal, and Weyuker 1994) and compilers (Wulf 1981; Su and Yan 2011). Programming languages consist of human-interpretable symbols and grammars capable of encoding an algorithm. Compilers take languages as inputs (low-information source) and expand these into machine-executable operations (high-information target). Starting with a thought that is expressed in a formal language—an algorithmic effective theory—a compiler extracts and maps tokenized elements of the language onto registers in a computer. These ultimately entrain the flow of fundamental particles, electrons, through semiconductors. Computers, programming languages, and compilers provide the constituent exemplars required to make sense of supervenience—they offer an engineered and highly practical proof of principle of the flow from low to high information. The key requirement for this kind of "downward causation" is that the initial whole needs to be far less than the sum of its final parts: *anti-emergence mechanisms*. The preoccupation with coarse-graining and information channels (both of which imply information loss to various degrees) has obscured the reality of programmed information expansion in engineered and living systems.

LANGUAGES

Programming languages are often described in terms of three major characteristics: (1) expressivity; (2) syntax; and (3) semantics. Expressivity captures the human-language interface with the computer-language in terms of naturalness, effort-requirements, and range of thought that can be translated into language. Syntax refers to the grammar of a language. Grammar is used in the sense of fixed rules describing interpretable sequences of tokens and operations, including the power of these rules (*Chomsky hierarchy*).[9] The semantics of a language describe its meaning in terms of the global function of a program (denotational semantics) or the local meaning of individual operations within the program (operational semantics). *Domain-specific languages* (**DSL**) impose restrictions on

---

[9] Chomsky hierarchy: A containment hierarchy of formal grammars—admissible strings produced by concatenating letters from an alphabet within a rule system—nested according to the complexity of their languages.

expressivity in order to align language syntax and semantics with the structure of their domain (Mathematica for symbolic manipulation, R for statistics, HTML for web pages, etc.).[10]

COMPILERS

The role of a compiler is to turn a high-level "source" language into register-level and memory-level "object" languages (targets). These objects are then optimized for a given hardware to produce machine-language operations. Compilers transform symbolic thoughts into instructions for the control of matter. Interpreters perform the same basic operations according to a different real-time model.

Compilation proceeds in distinct phases: (1) lexical analysis; (2) syntax analysis; (3) global optimization; (4) code generation; and (5) local optimization. Phases (1–3) can be thought of as the front-end of the computation process since they are at furthest remove from the details of the hardware. Phases (3–5) are back-end and highly machine-dependent.

☞ *Front-End Logic.* Lexical analysis discovers the boundaries of words in order to build up a symbol table and establish data types. Syntax analysis inputs sequences of tokens and constructs syntax trees corresponding to sequence of operations in target machines. Trees can be optimized to

---

[10] The most expressive computer language might be assumed to be a programming language that resembles spoken language, in which case programming would involve speaking a thought out loud. Spoken thoughts have the disadvantage of hiding various assumptions from the programmer. Making all assumptions explicit, understanding how results follow from these, and how ideas might be combined all provide important justifications for the choice of programming language: functional languages (composition of functions without regard for state and execution flow—highly abstract and efficient), procedural languages (execution of modular algorithms realized through concatenated subroutines—robust and generalizable), and imperative languages (direct control of execution flow and state changes—compact and problem-specific).

eliminate redundancies and improve the efficiency of control loops. Front-end logic is associated with distinct computational machines. Lexical analysis makes use of finite state machines and regular expressions. Syntax analysis implements a context-free grammar using a deterministic push-down machine.

☞ *Back-End Machine.* Tokens are converted into sequences of instructions that are expressed in a suitable assembly language, including operations loading into individual registers, into a global accumulator, or saving to memory. Tokens are scanned multiple times to ensure that conditional jumps are appropriately coded. The tokens are collected in a look-up table that assigns each a correct machine address.

☞ *Back-End Matter.* Each processor chip includes microcode, typically in the form of a ROM chip, that reads the opcode or machine code connecting to control lines that send read/write signals to transistors.

THE LOGICAL CONTROL OF MATTER

Supervenience in a complex system closely resembles compilation. For example, cells possess dedicated receptors that "read" and error-correct ligands binding at the cell surface. These activate a multiplicity of signaling pathways that culminate in differential gene expression in a context-sensitive way. Different gene regulatory networks (back-end) make use of the same upstream signaling pathways to translate extracellular inputs into "tokenized" activators of gene expression (front-end). The same logic can be extended to sensory activation and perception at the level of the whole nervous system: electromagnetic waves focused by the lens induce electrochemical activation in the retina, and signals propagate through the optic nerve and tracts, until they eventually converge and expand throughout visual cortex.

Computer languages are a means of expressing symbolic effective theories. Through compilation these languages are transformed from minimal sequences of signals into maximal states of material activation. Coordinated and collective modes of causal interaction at the material level are derived—top-down—from effectively causal algorithms.

The situation is of course materially different in biology and culture, at least as regards the logical sequence of compilation-like transformations, from inputs to tokens to registers. And the origin of these *organic generalized compiler*s is of course a complicated matter that needs to be described through an organic or cultural evolutionary process. An *organic-interpreter-model* (line by line tokenization and parsing) seems like a better fit to epigenetic modification and short-term memory.

The supposed paradox of low-dimensional inputs entraining high-dimensional outputs is resolved by compilers and interpreters. And top-down causality in the complex world arises from evolved pathways of information expansion—"many its from few bits."

Bibliography

Ahmed, A. 2010. *Wittgenstein's Philosophical Investigations: A Reader's Guide.* London, UK: Continuum.

Amari, S.-I. 1998. "Natural Gradient Works Efficiently in Learning." *Neural Computation* 10 (2): 251–276. https://doi.org/10.1162/089976698300017746.

Anderson, P. W. 1972. "More Is Different." *Science* 177 (4047): 393–396. https://doi.org/10.1126/science.177.4047.393.

Arthur, W. B. 1994. "Inductive Reasoning and Bounded Rationality." *American Economic Review* 84 (2): 406–411.

Babbage, C. 1832. *On the Economy of Machinery and Manufactures.* London, UK: Charles Knight.

—. 1837. *The Ninth Bridgewater Treatise: A Fragment.* London, UK: John Murray.

—. 1864. *Passages from the Life of a Philosopher.* London, UK: Longman.

—. 1982. "On the Mathematical Powers of the Calculating Engine." In *The Origins of Digital Computers*, 19–54. Berlin, Germany: Springer.

Bennett, C. H. 1973. "Logical Reversibility of Computation." *IBM Journal of Research and Development* 17 (6): 525–532. https://doi.org/10.1147/rd.176.0525.

Boltzmann, L. 1973. *The Boltzmann Equation: Theory and Applications.* Edited by E. G. D. Cohen and W. Thirring. New York, NY: Springer-Verlag Wien. https://doi.org/10.1007/978-3-7091-8336-6.

—. 1974. "On the Question of the Objective Existence of Processes in Inanimate Nature." In *Theoretical Physics and Philosophical Problems*, 57–76. Dordrecht, Netherlands: Springer.

—. 2012. *Theoretical Physics and Philosophical Problems: Selected Writings.* Berlin, Germany: Springer Science & Business Media.

Boole, G. 1847. *The Mathematical Analysis of Logic.* Cambridge, UK: Philosophical Library.

—. 1854. *An Investigation of the Laws of Thought on which are Founded the Mathematical Theories of Logic and Probabilities.* London, UK: Walton & Maberly.

—. 1859. *A Treatise on Differential Equations.* Cambridge, UK: Macmillan.

—. *Letter to Charles Babbage (October 15, 1862).* Add. MS 37198, no. 414. British Library, London, UK.

Bowler, P. J. 2005. "Revisiting the Eclipse of Darwinism." *Journal of the History of Biology* 38 (1): 19–32. https://doi.org/10.1007/s10739-004-6507-0.

Brush, S. G. 2003. *The Kinetic Theory Of Gases: An Anthology of Classic Papers with Historical Commentary.* Edited by N. S. Hall. River Edge, NJ: World Scientific. https://doi.org/10.1142/p281.

Burchfield, J. D. 2009. *Lord Kelvin and the Age of the Earth.* Chicago, IL: University of Chicago Press.

Callebaut, W., and R. Pinxten. 2012. *Evolutionary Epistemology: A Multiparadigm Program.* Dordrecht, Netherlands: Springer Science & Business Media.

Carnot, S. 1824. *Reflections on the Motive Power of Fire, and on Machines Fitted to Develop That Power.* Paris, France: Chez Bachelier.

Ceruzzi, P. E. 2003. *A History of Modern Computing, Second Edition.* Cambridge, MA: MIT Press.

Chaitin, G. J. 1966. "On the Length of Programs for Computing Finite Binary Sequences." *Journal of the ACM* 13 (4): 547–569. https://doi.org/10.1145/321356.321363.

Chalmers, D. J. 2006. "Strong and Weak Emergence." In *The Re-Emergence of Emergence*, edited by P. Clayton and P. Davies, 244–256. Oxford, UK: Oxford University Press.

Chalmers, T. 1853. *On the Power, Wisdom, and Goodness of God: As Manifested in the Adaptation of External Nature, to the Moral and Intellectual Constitution of Man.* London, UK: Bohn.

Clausius, R. 1879. *The Mechanical Theory of Heat.* London, UK: Macmillan.

Cobham, A. 1965. "The Intrinsic Computational Difficulty of Functions." In *Logic, Methodology and Philosophy of Science: Proceedings of the 1964 International Congress (Studies in Logic and the Foundations of Mathematics)*, edited by Y. Bar-Hillel, 24–30. Amsterdam, Netherlands: North- Holland Publishing.

Conant, R. C., and W. R. Ashby. 1970. "Every Good Regulator of a System Must be a Model of that System." *International Journal of Systems Science* 1 (2): 89–97. https://doi.org/10.1080/00207727008920220.

Corradini, A., and T. O'Connor. 2010. *Emergence in Science and Philosophy*. Hoboken, NJ: Taylor & Francis.

Crutchfield, J. P. 1994. "The Calculi of Emergence: Computation, Dynamics and Induction." *Physica D* 75 (1): 11–54. https://doi.org/10.1016/0167-2789(94)90273-9.

Darwin, C. 1859. *On the Origin of Species*. London, UK: John Murray.

—. 1861. *On the Origin of Species*. 3rd ed. London, UK: John Murray.

—. 1871. *The Descent of Man, and Selection in Relation to Sex*. Vol. 1. New York, NY: Appleton.

—. 1876. *The Movements and Habits of Climbing Plants*. 2nd ed. New York, NY: Appleton.

—. [1876] 1958. *The Autobiography of Charles Darwin: 1809-1882*. Edited by N. Barlow. Vol. 29. New York, NY: Collins.

Darwin, C., and A. Wallace. 1858. "On the Tendency of Species to form Varieties; and on the Perpetuation of Varieties and Species by Natural Means of Selection." *Zoological Journal of the Linnean Society* 3 (9): 45–62. https://doi.org/10.1111/j.1096-3642.1858.tb02500.x.

Darwin, F., and A. C. Seward. 1903. *More Letters of Charles Darwin*. Vol. 2. New York, NY: Appleton. Darwin Correspondence Project. 1869. Letter no. 6585. https://www.darwinproject.ac.uk/letter/?docId=letters/DCP-LETT-6585.xml.

—. 1878. *Letter no. 11729*. https://www.darwinproject.ac.uk/letter/?docId=letters/DCP- LETT- 11729.xml.

Davis, M., R. Sigal, and E. J. Weyuker. 1994. *Computability, Complexity, and Languages: Fundamentals of Theoretical Computer Science.* Amsterdam, Netherlands: Elsevier.

Dennett, D. 1995. *Darwin's Dangerous Idea: Evolution and the Meanings of Life.* New York, NY: Simon & Schuster.

Dilthey, W., and F. Jameson. 1972. "The Rise of Hermeneutics." *New Literary History* 3 (2): 229–244. https://doi.org/10.2307/468313.

Eigen, M., and R. Winkler. 1993. *Laws of the Game: How the Principles of Nature Govern Chance.* Princeton, NJ: Princeton University Press.

Farley, J. 1974. "The Initial Reactions of French Biologists to Darwin's 'Origin of Species'." *Journal of the History of Biology* 7 (2): 275–300.

Farmer, J. D. 1990. "A Rosetta Stone for Connectionism." *Physica D* 42 (1): 153–187. https://doi.org/10.1016/0167-2789(90)90072-W.

Feynman, R. P., R. B. Leighton, and M. Sands. 1963. *The Feynman Lectures on Physics.* Reading, MA: Addison-Wesley.

Forrest, S., A. S. Perelson, L. Allen, and R. Cherukuri. 1994. "Self–Nonself Discrimination in a Computer." In *Proceedings of 1994 IEEE Computer Society Symposium on Research in Security and Privacy,* 202–212. IEEE. https://doi.org/10.1109/RISP.1994.296580.

Galilei, G. 2001. *Dialogue Concerning the Two Chief World Systems, Ptolemaic and Copernican.* Edited by S. J. Gould. Translated by S. Drake. New York, NY: Modern Library.

Gell-Mann, M. 1994. *The Quark and the Jaguar: Adventures in the Simple and the Complex.* New York, NY: W. H. Freeman and Company.

Gell-Mann, M., and J. B. Hartle. 1994. *Equivalent Sets of Histories and Multiple Quasiclassical Realms.* arXiv: gr-qc/9404013 [gr-qc].

Gibb, S., R. F. Hendry, and T. Lancaster. 2019. *The Routledge Handbook of Emergence.* Abingdon, UK: Routledge.

Gibbs, J. W. 1875. "On the Equilibrium of Heterogeneous Substances." *Transactions of the Connecticut Academy of Arts and Sciences* 3:108–248.

—. 1902. *Elementary Principles in Statistical Mechanics.* New York, NY: Charles Scribner's Sons.

Gleick, J. [1987] 2008. *Chaos: Making a New Science.* New York, NY: Penguin.

Gribbin, J. 2004. *Deep Simplicity: Bringing Order to Chaos and Complexity.* New York, NY: Random House.

Guerra, C., M. Capitelli, and S. Longo. 2012. "The Role of Paradigms in Science: A Historical Perspective." In *Paradigms in Theory Construction*, edited by L. L'Abate, 19–30. New York, NY: Springer New York.

Hadamard, J. 1945. *The Psychology of Invention in the Mathematical Field.* New York, NY: Princeton University Press.

Haken, H. 1977. *Synergetics.* Berlin: Springer-Verlag.

Heilbron, J. L. 2010. *Galileo.* Oxford, UK: Oxford University Press.

Highfield, R., and P. Coveney. 1995. *Frontiers of Complexity.* New York, NY: Fawcett Columbine.

Hofstadter, D. R. 1979. *Gödel, Escher, Bach: An Eternal Golden Braid.* New York, NY: Basic Books.

Holland, J., and E. Domingo. 1998. "Origin and Evolution of Viruses." *Virus Genes* 16 (1): 13–21.

Holland, J. H. 1962. "Outline for a Logical Theory of Adaptive Systems." *Journal of the ACM* (New York, NY) 9 (3): 297–314. https://doi.org/10.1145/321127.321128.

—. 1995. *Hidden Order: How Adaptation Builds Complexity.* Cambridge, MA: Perseus Books.

—. 2014. *Complexity: A Very Short Introduction.* Oxford, UK: Oxford University Press.

Holland, J. H., and J. H. Miller. 1991. "Artificial Adaptive Agents in Economic Theory." *The American Economic Review* 81 (2): 365–370.

Holmes, K. V. 1999. "Coronaviruses (Coronaviridae)." *Encyclopedia of Virology* (San Diego, CA), 291.

Hopfield, J. J. 1982. "Neural Networks and Physical Systems with Emergent Collective Computational Abilities." *Proceedings of the National Academy of Sciences* 79 (8): 2554–2558. https://doi.org/10.1073/pnas.79.8.2554.

Jensen, H. J. 2022. *Complexity Science: The Study of Emergence*. Cambridge, UK: Cambridge University Press.

Jin, H. 2023. "The History, Current Applications and Future of Integrated Circuit." *Highlights in Science, Engineering and Technology*. https://doi.org/10.54097/hset.v31i.5146.

Johnson, N. 2009. *Simply Complexity: A Clear Guide to Complexity Theory*. Oxford, UK: Oneworld.

Kálmán, R. E. 1960. "Contributions to the Theory of Optimal Control." *Boletín de la Sociedad Matemática Mexicana* 5:102–119.

Karp, R. M. 1972. "Reducibility among Combinatorial Problems." In *Complexity of Computer Computations*, edited by R. E. Miller and J. W. Thatcher, 85–103. New York, NY: Plenum Press.

Kauffman, S. A. 1993. *The Origins of Order: Self-organization and Selection in Evolution*. New York, NY: Oxford University Press. https://doi.org/10.1093/oso/9780195079517.001.0001.

Kierkegaard, S. 1992. *Concluding Unscientific Postscript to Philosophical Fragments, A Mimical-Pathetic-Dialectical Compilation an Existential Contribution*. Translated by H. V. Wong and E. H. Wong. Vol. I. Princeton, NJ: Princeton University Press.

Kolmogorov, A. N. 1968. "Three Approaches to the Quantitative Definition of Information." *International Journal of Computer Mathematics* 2 (1–4): 157–168.

Krakauer, D. 2023. "Unifying Complexity Science and Machine Learning." *Frontiers in Complex Systems* 1. https://doi.org/10.3389/fcpxs.2023.1235202.

Kuhn, T. S. 2000. *The Road Since Structure: Philosophical Essays, 1970–1993, with an Autobiographical Interview*. Chicago, IL: University of Chicago Press.

—. [1962] 2012. *The Structure of Scientific Revolutions*. Chicago, IL: University of Chicago Press.

Ladyman, J., and K. Wiesner. 2020. *What Is a Complex System?* New Haven, CT: Yale University Press.

Landau, L. 2008. "On the Theory of Phase Transitions." Originally published in *Zh. Eksp. Teor. Fiz.* 7, pp. 19–32 (1937), *Ukrainian Journal of Physics* 53:25–35.

Landauer, R. 1961. "Irreversibility and Heat Generation in the Computing Process." *IBM Journal of Research and Development* 5 (3): 183–191. https://doi.org/10.1147/rd.53.0183.

Langton, C. G. 1986. "Studying Artificial Life with Cellular Automata." *Physica D* 22 (1): 120–149. https://doi.org/10.1016/0167-2789(86)90237-X.

Laughlin, R. B., D. Pines, J. Schmalian, B. P. Stojković, and P. Wolynes. 2000. "The Middle Way." *Proceedings of the National Academy of Sciences* 97 (1): 32–37. https://doi.org/10.1073/pnas.97.1.32.

Levin, S. 1999. *Fragile Dominion: Complexity and the Commons.* Reading, MA: Basic Books.

Lorenz, E. N. 1963. "Deterministic Nonperiodic Flow." *Journal of the Atmospheric Sciences* 20 (2): 130–141. https://doi.org/10.1175/1520-0469(1963)020<0130: DNF>2.0.CO;2.

Lyell, C. 1842. *Principles of Geology.* Boston, MA: Hilliard, Gray & Company.

Marr, D. 2010. *Vision: A Computational Investigation into the Human Representation and Processing of Visual Information.* Cambridge, MA: MIT Press. https://doi .org/10.7551/ mitpress/9780262514620.001.0001.

Mason, R. 1994. *Cambridge Minds.* Cambridge, UK: Cambridge University Press.

Maxwell, J. C. 1868. "I. On Governors." P*roceedings of the Royal Society of London* 16:270–283. https://doi. org/10.1098/rspl.1867.0055.

——. 1871. *Theory of Heat.* London, UK: Longmans, Green and Company.

——. 1872. "Molecules." In *The Scientific Papers of James Clerk Maxwell,* edited by W. D. Niven, vol. 2. Cambridge, UK: Cambridge University Press.

—. 1881. *An Elementary Treatise on Electricity*. Oxford, UK: Clarendon.

—. 1990. *The Scientific Letters and Papers of James Clerk Maxwell: Volume 2, 1862-1873*. Cambridge, UK: CUP Archive.

McCulloch, W. S., and W. Pitts. 1943. "A Logical Calculus of the Ideas Immanent in Nervous Activity." *Bulletin of Mathematical Biology* 5 (4): 115–133. https://doi.org/10.1007/BF02478259.

Menabrea, L. F., and A. Lovelace. 1843. *Sketch of the Analytical Engine Invented by Charles Babbage, Esq.* London, UK: Taylor & Francis.

Mendel, G. [1866]1925. *Experiments in Plant Hybridization*. Cambridge, MA: Harvard University Press.

Miller, J. G. 1955. "Toward a General Theory for the Behavioral Sciences." *American Psychologist* 10 (9): 513–531.

Miller, J. H. 2015. *A Crude Look at the Whole: The Science of Complex Systems in Business, Life, and Society*. New York, NY: Basic Books.

Minsky, M. 1961. "Steps Toward Artificial Intelligence." *Proceedings of the IRE* 49 (1): 8–30. https://doi. org/10.1109/JRPROC.1961.287775.

Mitchell, M. 2009. *Complexity: A Guided Tour*. Oxford, UK: Oxford University Press.

Mitchell, M., P. Hraber, and J. P. Crutchfield. 1993. "Revisiting the Edge of Chaos: Evolving Cellular Automata to Perform Computations." *Complex Systems* 7:89–130.

Miyazaki, J. 2013. *Pattern Formations and Oscillatory Phenomena: 2. Belousov–Zhabotinsky Reaction*. Amsterdam, Netherlands: Elsevier Inc.

Morowitz, H. J. 2002. *The Emergence of Everything: How the World Became Complex*. New York, NY: Oxford University Press.

Newton, I. 1934. *Sir Isaac Newton's Mathematical Principles of Natural Philosophy and His System of the World*. Translated by A. Motte. Berkeley, CA: University of California Press.

Page, S. E. 2010. *Diversity and Complexity*. New Jersey, NJ: Princeton University Press.

Palmer, R. E. 1969. *Hermeneutics: Interpretation Theory in Schleiermacher, Dilthey, Heidegger, and Gadamer.* Evanston, IL: Northwestern University Press.

Pederson, T. 2020. "The Double Helix: "Photo 51" Revisited." *The FASEB Journal* 34 (2): 1923–1927. https://doi.org/10.1096/fj.202000119.

Penrose, R. 2004. *The Road to Reality: A Complete Guide to the Laws of the Universe.* New York, NY: Knopf, Borzoi.

Peters, T. F. 1996. *Building the Nineteenth Century.* Cambridge, MA: MIT Press.

Piccinini, G., and A. Scarantino. 2011. "Information Processing, Computation, and Cognition." *Journal of Biological Physics* 37 (1): 1–38. https://doi.org/10.1007/s10867-010-9195-3.

Poincaré, H. 2017. The Three-Body Problem and the Equations of Dynamics. Translated by B. D. Popp. Cham, Switzerland: Springer International Publishing. https://doi.org/10.1007/978-3-319-52899-1.

—. [1903]2017. Science and Method. Translated by F. Maitland. New York, NY: Dover.

Prigogine, I., and P. M. Allen. 1982. "The Challenge of Complexity." In *Self-Organization and Dissipative Structures,* edited by W. C. Schieve and P. M. Allen, 1–39. Austin, TX: University of Texas Press.

Prigogine, I., and I. Stengers. 2018. *Order Out of Chaos: Man's New Dialogue with Nature.* London, UK: Verso Books.

Pylyshyn, Z. W. 1984. *Computation and Cognition: Toward a Foundation for Cognitive Science.* Cambridge, MA: MIT Press.

Radick, G. 2023. *Disputed Inheritance: The Battle over Mendel and the Future of Biology.* Chicago, IL: University of Chicago Press.

Rashevsky, N. 1956. "The Geometrization of Biology." *Bulletin of Mathematical Biophysics* 18 (1): 31–56. https://doi.org/10.1007/BF02477842.

Richards, R. J. 2013. "The German Reception of Darwin's Theory, 1860–1945." In *The Cambridge Encylopedia of Darwin and Evolutionary Thought,* edited by M. Ruse, 235–242. Cambridge, UK: Cambridge University Press.

Rosenblueth, A., N. Wiener, and J. Bigelow. 1943. "Behavior, Purpose and Teleology." *Philosophy of Science* 10 (1): 18–24.

Sammet, J. E. 1969. *Programming Languages: History and Fundamentals.* Englewood Cliffs, NJ: Prentice-Hall.

Schelling, T. C. 1971. "Dynamic Models of Segregation." *Journal of Mathematical Sociology* 1 (2): 143–86. https://doi.org/10.1080/0022250X.1971.9989794.

Schrödinger, E. 1944. *What is Life? The Physical Aspect of the Living Cell.* Cambridge, UK: Cambridge University Press.

Shannon, C. E., and W. Weaver. 1949. *The Mathematical Theory of Communication.* Urbana, IL: University of Illinois Press.

Shapiro, J. A. 2009. "Letting *Escherichia coli* Teach Me About Genome Engineering." *Genetics* 183 (4): 1205– 1214. https://doi.org/10.1534/genetics.109.110007.

Simon, H. A. 1962. "The Architecture of Complexity." *Proceedings of the American Philosophical Society* 106 (6): 467–482.

—. 1977. "The Organization of Complex Systems." In *Models of Discovery: And Other Topics in the Methods of Science*, edited by H. A. Simon, 245–261. Springer Netherlands.

Snyder, L. 2011. *The Philosophical Breakfast Club: Four Remarkable Friends*. New York, NY: Broadway Books.

Solomonoff, R. J. 1964. "A Formal Theory of Inductive Inference. Part I." *Information and Control* 7 (1): 1–22. https://doi.org/10.1016/S0019-9958(64)90223-2.

Stein, D. L., and C. M. Newman. 2013. *Spin Glasses and Complexity.* Princeton, UK: Princeton University Press.

Stent, G. 2002. *Prematurity in Scientific Discovery: On Resistance and Neglect.* Edited by E. B. Hook. Berkeley, CA: University of California Press.

Su, Y., and S. Y. Yan. 2011. *Principles of Compilers.* Berlin, Germany: Springer.

Szilárd, L. [1929]1964. "On the Decrease of Entropy in a Thermodynamic System by The Intervention of Intelligent Beings." *Behavorial Science* 9 (4): 301–310. https://doi.org/10.1002/bs.3830090402.

Taylor, R., ed. 1843. *Scientific Memoirs, Selected from the Transactions of Foreign Academies of Science and Learned Societies.* Translated by A. Lovelace. "Translator's Notes to Mr. Menabrea's Memoir." London, UK: Longman.

Thompson, D. W. 1917. *On Growth and Form*. Cambridge, UK: Cambridge University Press.

Thomson, W. 1874. "Kinetic Theory of the Dissipation of Energy." *Nature* 9:441–444. https://doi.org/10.1038/009441c0.

Thurner, S., R. Hanel, and P. Klimek. 2018. *Introduction to the Theory of Complex Systems*. Oxford, UK: Oxford University Press.

Topham, J. R. 2022. *Reading the Book of Nature: How Eight Best Sellers Reconnected Christianity and the Sciences on the Eve of the Victorian Age.* Chicago, IL: University of Chicago Press.

Turing, A. M. [1950]2012. "Computing Machinery and Intelligence." In *The Essential Turing: the Ideas That Gave Birth to the Computer* Age, 433–464. Oxford, UK: Oxford University Press.

Ulam, S. 1962. "On Some Mathematical Problems Connected with Patterns of Growth in ws." In *Mathematical Problems in the Biological Sciences,* edited by R. Bellman, 215–224. American Mathematical Society. https://doi.org/10.1090/psapm/014/9947.

Varela, F. G., H. R. Maturana, and R. Uribe. 1974. "Autopoiesis: The Organization of Living Systems, its Characterization and a Model." *Biosystems* 5 (4): 187–196. https://doi.org/10.1016/0303-2647(74)90031-8.

von Bertalanffy, L. 1950. "The Theory of Open Systems in Physics and Biology." *Science* 111 (2872): 23–29.

von Foerster, H. 2003. "Notes on an Epistemology for Living Things." In *Understanding Understanding: Essays on Cybernetics and Cognition*, 247–259. New York, NY: Springer New York.

von Hayek, F. A. 1945. "The Use of Knowledge in Society." *American Economic Review* 35:519–530.

von Neumann, J. 1966. *Theory of Self-Reproducing Automata.* Edited by A.W. Burks. Urbana, IL: University of Illinois Press. https://doi.org/10.5555/1102024.

Waldrop, M. M. 1992. *Complexity: The Emerging Science at the Edge of Order and Chaos.* New York, NY: Simon & Schuster.

Wallace, A. R. 1889. *Darwinism: An Exposition of the Theory of Natural Selection with Someof Its Applications.* 2nd ed. London, UK: Macmillan.

Weaver, W. 1948. "Science and Complexity." *American Scientist* 36 (4): 536–544.

West, G. B. 2018. Scale: *The Universal Laws of Life, Growth, and Death in Organisms, Cities, and Companies.* New York, NY: Penguin.

West, G. B., and J. H. Brown. 2005. "The Origin of Allometric Scaling Laws in Biology from Genomes to Ecosystems: Towards a Quantitative Unifying Theory of Biological Structure and Organization." *The Journal of Experimental Biology* 208 (9): 1575–1592. https://doi.org/10.1242/jeb.01589.

Wheeler, J. A. 1990. "Information, Physics, Quantum: The Search for Links." In *Complexity, Entropy, and the Physics of Information,* edited by W. H. Zurek, 3–28. Reading, MA: Addison–Wesley.

Whewell, W. 1833. *Astronomy and General Physics Considered with Reference to Natural Theology.* London, UK: Pickering.

Wiener, N. 1948. *Cybernetics: or Control and Communication in the Animal and the Machine.* New York, NY: John Wiley and Sons.

Wigner, E. 1964. "Events, Laws of Nature, and Invariance Principles." *Science* 145 (3636): 995–999. https://doi.org/10.1126/science.145.3636.995.

Wittgenstein, L. [1953]2009. *Philosophical Investigations.* 4th ed. Edited by P. M. S. Hacker and J. Schulte. Translated by G. E. M. Anscombe. Malden, WI: Wiley-Blackwell.

Wolfram, S. 1984a. "Cellular Automata as Models of Complexity." *Nature* 311 (5985): 419–424. https ://doi.org/10.1038/31141920.

—. 1984b. "Universality and Complexity in Cellular Automata." *Physica D* 10 (1): 1–35. https://doi. org/10.1016/0167-2789(84)90245-8.

—. 2018. *A New Kind of Science*. Champaign, IL: Wolfram Media.

Wray, B. K. 2021. *Interpreting Kuhn: Critical Essays*. Cambridge, UK: Cambridge University Press.

Wulf, W. A. 1981. "Compilers and Computer Architecture." *Computer* 14 (07): 41–47. https://doi.org/10.1109/C-M.1981.220527.

Zenil, H. 2013. *A Computable Universe: Understanding and Exploring Nature as Computation*. River Edge, NJ: World Scientific.

How to Cite

When citing the standalone introduction to *Foundational Papers of Complexity Science*, please use the following approach:

BIBLIOGRAPHY:

Krakauer, David C. *The Complex World: An Introduction to the Foundations of Complexity Science.* Santa Fe, NM: SFI Press, 2024.

LATEX:

```
@book{Krakauer_FP_intro_2024,
        author = {Krakauer, David C.},
        title = {The Complex World: An
                Introduction to the Foundations
                of Complexity Science},
        year = {2024},
        publisher = {SFI Press},
        location = {Santa Fe, NM}
}
```

FOUNDATIONAL PAPERS: Volume 1

Available for purchase. Visit **www.sfipress.org** *to learn more.*

FOUNDATIONAL PAPERS: Volume 2

Available for purchase. Visit **www.sfipress.org** *to learn more.*

FOUNDATIONAL PAPERS: Volume 3

Available for purchase. Visit **www.sfipress.org** *to learn more.*

FOUNDATIONAL PAPERS: Volume 4

Available for purchase. Visit www.sfipress.org to learn more.

Artwork in this Volume

In the fifteenth and sixteenth centuries, geometry promised a key to all reality, rendering the cluttered world into logical abstraction through perspective.

For artist–mathematicians Piero della Francesca (c. 1415–1492) and Paolo Uccello (1397–1475), reality was best served on rails. Albrecht Dürer (1471–1528) placed regular polygons within melancholic studios to comfort failing scholars. And Johannes Kepler (1571–1630) suspended the delicate earth within a progressive series of expanding solids to harmonize the inconstant universe.

From within and without, Platonic order brought order to chaos. And yet for others, whose names have been lost or erased, these idealizations were little more than abstract games to amuse the fallen, and timeless exercises opposed to temporal reality.

In the Herzog August Bibliothek in Wolfenbüttel there is an anonymous manuscript. It depicts a fleeting world of birds and bugs interrogating in their distracted fashion the grandiose claims of solid geometry. A sparrow puzzles over the shadow of an inverted pyramid, a squirrel laces acorn shoots beside an octahedron, and a snail approaches a stellated polyhedron balanced by parakeet. Songbirds entrance diverse crystal groups into acrobatics.

In this parallel world life is suspicious of symmetry and treats Plato's ideal as decoration, invader, ornament, and nest. Vertices, lattices, grids, reticulations, and meshes are the degrading ornaments of Apollo's garden as invaded by the species of Dionysus.

To view all paintings in the original manuscript (Collection of geometric and perspective drawings executed in color, comprising 36 sheets: Cod. Guelf. 74.1 Aug. 2°; Heinemann No. 2708), please visit the Herzog August Bibliothek Wolfenbüttel: https://diglib.hab.de/wdb.php?dir=mss/74-1-aug-2f&distype=thumbs

Author

David C. Krakauer is the President and William H. Miller Professor of Complex Systems at the Santa Fe Institute. His research explores the evolution of intelligence and stupidity on Earth. This includes studying the evolution of genetic, neural, linguistic, social, and cultural mechanisms supporting memory and information processing, and exploring their shared properties. He served as the founding director of the Wisconsin Institutes for Discovery, codirector of the Center for Complexity and Collective Computation, and professor of mathematical genetics, all at the University of Wisconsin, Madison. He has been a visiting fellow at the Genomics Frontiers Institute at the University of Pennsylvania, a Sage Fellow at the Sage Center for the Study of the Mind at the University of California, Santa Barbara, a longterm fellow of the Institute for Advanced Study, and visiting professor of evolution at Princeton University. In 2012, he was included in the *Wired Magazine* Smart List: Fifty People Who Will Change the World. In 2016, he was included in *Entrepreneur Magazine*'s list of visionary leaders advancing global research and business.

Krakauer was previously chair of faculty and a resident professor and external professor at the Santa Fe Institute. A graduate of the University of London where he earned degrees in biology and computer science, he received his D.Phil. in evolutionary theory from Oxford University in 1995 and continued there as a postdoctoral fellow.

ABOUT THE SANTA FE INSTITUTE

The Santa Fe Institute is the world headquarters for complexity science, operated as an independent, nonprofit research and education center located in Santa Fe, New Mexico. Our researchers endeavor to understand and unify the underlying, shared patterns in complex physical, biological, social, cultural, technological, and even possible astrobiological worlds. Our global research network of scholars spans borders, departments, and disciplines, bringing together curious minds steeped in rigorous logical, mathematical, and computational reasoning. As we reveal the unseen mechanisms and processes that shape these evolving worlds, we seek to use this understanding to promote the well-being of humankind and of life on Earth.

COLOPHON

The body copy for this book was set in EB Garamond, a typeface designed by Georg Duffner after the Ebenolff–Berner type specimen of 1592. Headings are in Kurier, created by Janusz M. Nowacki, based on typefaces by the Polish typographer Małgorzata Budyta, and Futura PT, based on Paul Renner's 1927 geometric sans-serif typeface. Footnotes and captions also incorporate variations of Futura.

The SFI Press complexity glyphs used throughout our books were designed by Brian Crandall Williams.

SANTA FE INSTITUTE
COMPLEXITY
GLYPHS

ZERO

ONE

TWO

THREE

FOUR

FIVE

SIX

SEVEN

EIGHT

NINE

-A-

-B-

-C-

-D-

-E-

-F-

-G-

-H-

-I-

-J-

-K-

-L-

-M-

-N-

-O-

-P-

-Q-

-R-

-S-

-T-

-U-

-V-

-W-

-X-

-Y-

-Z-

Notes ♡

Notes 6

Notes

Notes

Notes ⊣

Notes

Notes

SFI PR_SS

SCHOLARS SERIES

www.ingramcontent.com/pod-product-compliance
Lightning Source LLC
Chambersburg PA
CBHW041914190326
41458CB00024B/6259